Endorsements

If there was ever a call for energy innovation, *Renewed Energy* reminds us that we must act now. With the rising threat of greenhouse gas emissions, climate change is a global problem that cannot be ignored. The interviews in *Renewed Energy* must germinate beyond the pages and lead to action at research centers and street corners.

– Robert B. Weisenmiller
Chair, California Energy Commission

Renewed Energy is a wonderfully readable account of the successes and tribulations of the early years of cleantech, and also a call to arms for the revitalized next wave. Energy technology transformations are today ripe for the picking in ways that would have been inconceivable 15 years ago. These stories will inspire and inform the next round of innovators, investors, and policymakers who advance those opportunities.

– Ellen Williams
Former Director, Advanced Research Projects Agency-Energy (ARPA-E) and Distinguished University Professor, University of Maryland

This book provides unique and important insights into the development of clean energy through the lens of private investors and federal U.S. energy support. It showcases a number of prominent figures and provides excellent insight into the groundbreaking leadership they have provided. A recommended read for those seeking to understand how clean energy has flourished and the expanding role it will play going forward.

– Dian Grueneich
Former Commissioner, California Public Utilities Commission

We need cleaner energy more than ever, but to achieve that future, we must learn from the first wave of cleantech efforts. *Renewed Energy* uncovers the history of clean energy's first decade with a level of understanding we haven't seen before. These powerful stories and first-hand insights—into what went well and what went wrong—are synthesized into powerful learnings for cleantech companies, investors, and policymakers alike.

– **Kenneth Davies**
Director of Renewable Energy, Microsoft

Renewed Energy brings together a wealth of authoritative voices to underscore the promise of clean energy innovation. Spanning titans of industry, government, and academia, the experts interviewed in the book grapple with the complexity of overhauling the global energy system and share insider insights into what it takes to commercialize new energy technologies and displace old ones. Despite the daunting challenge, however, the book is essentially optimistic. It reassures us that policymakers, firms, and investors can all draw on lessons from past experience to intelligently advance new technologies through scientific ingenuity, entrepreneurial pluck, and bipartisan policies.

– **Varun Sivaram**
Philip D. Reed Fellow for Science and Technology, Council on Foreign Relations and Author of *Taming the Sun: Innovations to Harness Solar Energy and Power the Planet*

Ernestine Fu, John Weyant, and Justin Bowersock have done warm and genuine interviews with thought leaders in energy, policy, and investing. The book provides a timely reflection on cleantech investing in the last decade, and shares important lessons learned. The authors make it clear that collaboration between policymakers, industry, and venture capital investors is the way to move forward in creating a sustainable and carbon-neutral economy.

– **Jiong Ma**
Partner, Braemar Energy Ventures

Reading *Renewed Energy* is like going to a cocktail party at an exclusive clean energy conference. First you're chatting with the policy architects of the last 20 years, asking what they are proudest of or would do differently; then you find yourself in candid conversations with private investors and entrepreneurs, who met the market head on to build clean energy businesses in a complex international landscape. Ultimately, you may leave with more questions than answers, but you're definitely satisfied that you've talked to all the right people.

– Hemai Parthasarathy
Chief Science Officer and Partner, Breakout Ventures

Renewed Energy is an astute, lucid and important study of the first wave of cleantech investments in the past decade that led to many disappointments. It is the first serious attempt to address two questions with serious implications for the future of the human race and our national competitiveness: What went wrong with the first wave of cleantech investments? How do we do this better next time? The book will not end the quest for those answers, but will certainly guide the readers to the right trail.

Prior media discussions and scholarly analyses have often dismissed the first wave of cleantech investments either as crazed mania, or as misguided efforts by traditional VC firms whose structures are unsuited for this industry. Both arguments are wrong. The cleantech investments were led by smart and accomplished people driven by rational intentions. And, despite early failures, many VC and PE firms are now succeeding and proving that cleantech can be both good and profitable.

The truth is more complex and nuanced. A newly evolving generation of entrepreneurs, investors, and policymakers certainly made mistakes, but those errors were severely compounded by a thicket of market frictions, ideological disputes, economic recession, and policy whiplashes. This book takes an intimate look at those years from divergent perspectives of many principal characters. What emerges is an analytical framework that provides a refreshing explanation of the past and an illuminating view to the future.

Renewed Energy brings clarity and cogency to one of the biggest issues of our time, and will act as a beacon of light to those trying to find superior, sustainable, and affordable solutions to clean energy, food, water, and waste. Please read this deeply engaging book and be prepared to do some serious rethinking!

– Praveen Sahay
Managing Director, WAVE Equity Partners

Renewed Energy: Insights for Clean Energy's Future

John Weyant, Ernestine Fu, and Justin Bowersock

Under no circumstances shall any of the information provided herein be construed as investment advice of any kind.

Renewed Energy:
Insights for Clean Energy's Future
John Weyant, Ernestine Fu, and Justin Bowersock

Published by Kauffman Fellows Press
855 El Camino Real, Suite 12, Palo Alto, CA 94301, United States
www.kauffmanfellows.org

ISBN-13: 978-1-939533-02-9
Printed in the United States of America

Editing and design by Together Editing & Design
www.togetherediting.com

First Edition: August 2018
9 8 7 6 5 4 3 2 1

We are grateful for the support of several research assistants and our editors who made this publication possible. They were instrumental in conducting research, transcribing interviews, writing and copyediting, and most importantly, providing incredibly wise counsel.

All profits received by the authors from the sale of the book will be donated to philanthropic organizations that accelerate our transition to a renewable energy economy.

Underlined names and terms are described in the Glossary at the end of the book; **names in bold** indicate that an interview with that person can be found elsewhere in the book.

Contents

Introduction: Energy's Three Separate Worlds

Energy today is three seemingly separate worlds—policy, technology investing, and industry. Each is constructed largely in ignorance of the other two, but all three are so interconnected that they are impossible to affect singly. While many have applied themselves to developing solutions to society's complex energy problems, progress has been limited. We believe the biggest gap in energy today is the lack of a holistic approach that marries technology progress and investment to industry and policy.

In this publication, we address that gap by providing the context necessary to begin to understand and unravel the complexities of energy, specifically greentech or cleantech. Our goal is to educate scientists, venture capitalists, politicians, entrepreneurs, and executives alike. We hope that insights into these three interconnected worlds will help drive necessary changes—and improvements—in the cleantech sector.

The three of us first met at Stanford University in 2012, and we initially did not think that casual meetings precipitated by our shared interests would result in anything more than a couple of chats over coffee or comparing notes on our research and observations. John Weyant has had the good fortune and opportunity to work with the best and brightest minds in government and the scientific community. Ernestine Fu sees a thousand companies a year, sits on private company governing boards, and advises many others. Justin Bowersock has seen the energy industry from perspectives ranging from Wall Street's commodity trading floor to the hard-hat sites of utility-scale renewable projects. Bringing these three perspectives together gave us an understanding of many of the levers and drivers of the industry over the past decade.

A few months and years into our conversations, we felt a shared call to action. We began to interview stakeholders and influencers in the energy industry, from policymakers to technology investors and founders. We did not know if the interviews would result in anything, but we realized that the future of cleantech required an engagement with an understanding of the industry's past—and that the synergy of diverse perspectives was the key to that understanding.

Our observations began to meld into clear, concise, and actionable ideas. As we realized the importance of our three-pronged approach, we knew we needed to foster understanding for a broad audience and share our thoughts and conversations. Using our holistic lens, we together could unravel the roles of technology investing, industry, and policy in cleantech's history and identify meaningful insights for the future.

This publication is about arming readers with enough knowledge to come to conclusions and discover insights on their own—and not just through our lenses. We incorporate an array of experts in each sector—some of the world's most experienced and knowledgeable people—who share their thoughts through the interviews we conducted and now present here.

For example, readers can compare and contrast the ideas of **Tom Baruch**, who led the early investment in Solyndra at CMEA Capital, with those of **Arun Majumdar**, who architected the original policy behind the U.S. Department of Energy's Advanced Research Projects Agency-Energy (ARPA-E). Readers can also gather insights from other global leaders such as Secretary **Steven Chu**, who led energy policy under the Obama administration, and **Artur Runge-Metzger**, who was the Chief Climate Negotiator for the European Commission.

The outcome of this approach is a powerful tool that furnishes more information than any of us alone could have brought to the table. Our hope is that readers not only enjoy these 11 interviews, but also gain the framework to better approach energy innovation in a positive, integrated, and comprehensive manner.

Today (2018) is an interesting time for clean energy. We conducted our interviews during a time when federal government policy under the Barack Obama administration was concerned with climate change and mitigating its effects. The Environmental Protection Agency's clean power plan sought to lower carbon emissions on a state-by-state basis. In signing the Paris agreement, countries committed to voluntary limits on future carbon emissions. As the new presidential administration has ended the U.S. clean power plan and decided to exit the Paris agreement, we believe these interviews are—now more than ever—critical to share.

We can align the stars for solving our world's energy problems, but it will take innovation, perseverance, leadership, and collaboration. Most importantly, we need cooperation and insights from policymakers, technology innovators and investors, and energy industry companies, together. Change doesn't happen overnight—and change starts small, beginning with conversations. Join us in reading our conversations and interviews with energy leaders. Together, we hope to effect positive changes in the energy sector and society.

Technology Investing and Industry Perspective

From a technology investing and industry perspective, the story of cleantech and renewable energy started decades ago, although mainstream media and the public only recently began to pay attention. Like the Information Age that came before it, cleantech was poised to be the next great technological movement with significant returns for investors. While many already know this movement's ill-fated outcome following its investment peak of $10.2 billion in 2008,[1] few understand the reasons.

Cleantech investing and IT investing are affected by the same broader market events, but they are fundamentally different. It is vital to understand where their stories diverge.

For IT investing, we begin with March 2000: the floor fell out from under the Nasdaq.[2] It was the worst possible time in recent decades to be an investor. Portfolios became catacombs of dead or dying companies that produced poor returns. An entire industry of venture capitalists was left adrift and directionless.

It wasn't until a decade later, in 2010, that new life was breathed back into investing and venture capital. Enter LinkedIn, the professional social network that was about to blow the doors off the IPO market[3] and reinvigorate a weary

[1] Frankfurt School–UNEP Collaborating Centre, *Global Trends in Renewable Energy Investment 2017*, fig. 48, http://fs-unep-centre.org/sites/default/files/publications/globaltrendsinrenewable energyinvestment2017.pdf.

[2] Ben Geier, "What Did We Learn from the Dotcom Stock Bubble of 2000?" *Time*, 12 March 2015, http://time.com/3741681/2000-dotcom-stock-bust/.

[3] Julianne Pepitone, "LinkedIn Stock More Than Doubles in IPO," *CNN Money*, 19 May 2011, http://money.cnn.com/2011/05/19/technology/linkedin_IPO/index.htm.

investment community—and on its coattails, Facebook,[4] the soon-to-be tech giant that would redefine what being social meant.

Together, the two ushered in a new era in tech, named "Internet 2.0." LinkedIn IPO'd in 2011 at $45 per share and would grow as high as $269 per share, ultimately being purchased 5 years later by Microsoft for $196 per share.[5] Similarly, Facebook debuted at $38 per share, has grown steadily year over year, and as of this writing, is sitting at an enviable $187 per share.[6]

The story of cleantech and renewable energies shared a similar beginning as the dotcom bubble. Cleantech followed the motif of copious investments made with exuberance for a new and exciting market—one that investors could feel socially good about. The two protagonists who spurred this parallel tale came together in storybook fashion: technology (John Doerr) and government (Al Gore).

In 2006, Al Gore launched the documentary *An Inconvenient Truth*, an independent film that became the 10th-largest grossing documentary.[7] His film served as a rallying cry for everyone, everywhere, to take up the cause of mitigating global climate change, pollution, and energy problems.

The first to heed that cry was John Doerr. In his stirring 2007 TED Talk, "Salvation (and Profit) in Greentech," Doerr delivered a riveting address packed with both emotion and unfortunate facts.[8] He integrated important people and trends, from Gore's documentary to Walmart's then-underway initiative to go green across all business operations.[9] Doerr ended with this big request: "So, our call to action—my call to you—is for you to make going green your next big thing, your gig."[10]

Leading by example, Doerr made "going green" his next gig, steering his venture capital firm—Kleiner, Perkins, Caufield and Byers (KPCB)—headlong

[4] Alex Wilhelm, "A Look Back in IPO: Facebook's Trailing Profit and Mobile Intrigue," *TechCrunch*, 22 August 2017, https://techcrunch.com/2017/08/22/a-look-back-in-ipo-facebooks-trailing-profit-and-mobile-intrigue/.

[5] Google Finance, "LinkedIn Corp,." 9 January 2018, https://finance.google.com/finance?q=NYSE:LNKD.

[6] Google Finance, "Facebook Inc.," 9 January 2018, https://finance.google.com/finance?q=facebook.

[7] A.O. Scott, "Warning of Calamities and Hoping for a Change in 'An Inconvenient Truth,'" *The New York Times*, 24 May 2006, http://www.nytimes.com/movies/movie/342290/An-Inconvenient-Truth/awards.

[8] John Doerr, "Salvation (and Profit) in Greentech" (video), *TED*, March 2007, http://www.ted.com/talks/john_doerr_sees_salvation_and_profit_in_greentech?language=en.

[9] Mark Gunther, "The Green Machine," *Fortune*, 31 July 2006, http://archive.fortune.com/magazines/fortune/fortune_archive/2006/08/07/8382593/index.htm; Wal-Mart has since announced that 81% of its waste material in the United States is being diverted from landfills. See "Wal-Mart's Green Initiative: Status Report," *The Wall Street Journal*, 8 April 2014, http://www.wsj.com/articles/SB10001424052702304432604579473453226974252.

[10] Doerr, "Salvation," 15:36.

into cleantech venture funding. Just a few short months later, Gore joined Doerr at KPCB as a partner.[11]

Other investors followed suit, most notably Vinod Khosla, with his investments in green companies ranging from new types of batteries to clean cement and greener glass. Maintaining the focus on public interest, Khosla participated in high-profile public interviews discussing his commitment to make money with cleantech.[12] The mainstream media followed suit, with *Forbes* naming Doerr as the #1 venture capitalist in the world on the 2008 Midas List of top technology investors, and also rewarding new entrant Khosla with a top-100 spot.[13]

A surge in cleantech investments paralleled the increase in public interest, from $2.9 billion in 2006[14] to an astronomical $4.1 billion two years later.[15] At that point, however, the stories of IT investing and cleantech investing diverge.

Despite the surge in investments and the optimism of Gore, Doerr, and Khosla, 2008 may well have been the end of the party. Private venture capital investment in cleantech decreased 37% in 2009, to $2.6 billion.[16]

While the 2008 housing bubble and financial crisis played a role in this decline, the deeper and more fundamental issue was the lack of later-stage support for the cleantech companies funded in the first wave of financing.[17] Traditional hardware, software, and Internet technology companies have a mature ecosystem at every investment stage, from seed through growth to IPO, whereas at the end of 2008, there was simply no institutional support for companies like Solyndra, Fisker, and Tesla to continue raising expendable capital.

[11] Matt Marshall, "Al Gore Joins Kleiner Perkins as a Partner – To Push Green Investments," *VentureBeat*, 12 November 2007, http://venturebeat.com/2007/11/12/al-gore-joins-kleiner-perkins-as-a-partner-to-push-green-investments/.

[12] Anupreeta Das, "Khosla Looking for Clean Energy 'Black Swans,'" *Reuters*, 7 October 2008, https://www.reuters.com/article/us-summit-khosla/khosla-looking-for-clean-energy-black-swans-idUSTRE49719S20081008.

[13] Forbes, "The Midas List," 24 January 2008, http://www.forbes.com/lists/2008/99/biz_08midas_The-Midas-List_Rank.html.

[14] James Stack, *Cleantech Venture Capital: How Public Policy Has Stimulated Private Investment* (Environmental Entrepreneurs and Cleantech Network LLC, 15 May 2007), 8, https://members.e2.org/ext/doc/CleantechReport2007.pdf.

[15] Shikar Gosh and Ramana Nada, "Venture Capital Investment in the Clean Energy Sector" (working paper), Harvard Business School 11-020, 1 August 2010, 8, http://core.ac.uk/download/pdf/6698655.pdf.

[16] Ernst & Young LLP, "Venture Capital 2009 Investments in Cleantech Fall 50% to $2.6 Billion as Investors Shift Focus to Energy Efficiency" (news release), 8 February 2010, para. 1, http://www.prnewswire.com/news-releases/venture-capital-2009-investments-in-cleantech-fall-50-to-26-billion-as-investors-shift-focus-to-energy-efficiency-83790562.html.

[17] Devashree Saha and Mark Muro, "Cleantech Venture Capital: Continued Declines and Narrow Geography Limit Prospects," *Brookings*, 16 May 2017, https://www.brookings.edu/research/cleantech-venture-capital-continued-declines-and-narrow-geography-limit-prospects/.

The limitations on cleantech companies' growth were addressed—but through politics and government spending rather than venture capital investing. After the 2008 presidential election, Barack Obama and the Democrats took complete control of the federal government for several years. This mandate opened the door to advance the party's traditional ideals, and clean energy fit into that agenda perfectly. The Department of Energy's $4 billion loan program was repurposed in 2009, allowing the DoE to start doling out loans and tax credits to privately funded companies like Solyndra and Fisker, filling the vacuum created by the 2009 recession and the sector's lack of growth-stage investors.[18] Across just two years, from 2009 to 2011, these investments totaled more than $35 billion.[19]

If there was a battle of "cleantech versus social" in investors' minds over the question of where to put their dollars, the loser was decided in 2011: Solyndra went bankrupt that August, just two months after LinkedIn went public. Solyndra had raised $1.6 billion[20] and failed completely,[21] while during the same timeframe LinkedIn had raised only approximately $100 million in additional capital[22] and took off with an IPO north of $10 billion.[23] Battery-maker A123, once viewed as among the most promising companies for electric car development, also went bankrupt in October 2012.[24]

What went wrong?

The first problem was that tech investors—including Doerr and Khosla—had attempted to create an ecosystem where none existed. It is ironic that a famous set of modern-day proverbs by the founder of Doerr's firm (KPCB), known as

[18] U.S. Department of Energy, Loan Programs Office (LPO), "History of the Loan Programs Office," https://energy.gov/lpo/timeline/history-loan-programs-office.

[19] Jeff Brady, "After Solyndra Loss, U.S. Energy Loan Program Turning a Profit," *NPR*, 13 November 2014, para. 2, http://www.npr.org/2014/11/13/363572151/after-solyndra-loss-u-s-energy-loan-program-turning-a-profit.

[20] Crunchbase, "Solyndra," https://www.crunchbase.com/organization/solyndra#/entity.

[21] Steven Church, Joe Schneider, and Linda Sandler, "Solyndra, Solar-Panel Maker in California, Seeks Bankruptcy," *Bloomberg*, 6 September 2011, https://www.bloomberg.com/news/articles/2011-09-06/solyndra-solar-panel-maker-in-california-files-for-bankruptcy-protection.

[22] Crunchbase, "LinkedIn," https://www.crunchbase.com/organization/linkedin#/entity.

[23] Clare Baldwin and Alina Selyukh, "LinkedIn Share Price More Than Doubles in NYSE Debut" *Reuters*, 18 May 2011, http://www.reuters.com/article/2011/05/19/us-linkedin-ipo-risks-idUSTRE74H0TL20110519; Pepitone, "LinkedIn Stock More Than Doubles."

[24] Bill Vlasic and Matthew L. Wald, "Maker of Batteries Files for Bankruptcy," *The New York Times*, 16 October 2012, http://www.nytimes.com/2012/10/17/business/battery-maker-a123-systems-files-for-bankruptcy.html.

"Kleiner's Laws," states: "It's easier to get a piece of an existing market than to create a new one."[25] Doerr and Khosla had no other option, though—there was no existing ecosystem in which cleantech startups could mature into stable businesses, and many cleantech companies were attempting to create entirely new energy markets that were to compete against the well-established traditional energy sector.

The problem was that the macroeconomic trends of energy are capricious and ungovernable. For example, when the price of polycrystalline silicon (polysilicon) climbed in the mid-2000s, so did the cost of traditional solar panel cells. While this paved the way for Solyndra's precipitous rise to prominence, it also left the company vulnerable to price reversal. As polysilicon costs fell to all-time lows in 2009, Solyndra's alternative technology could not possibly compete on price, no matter how many DoE loans they secured.[26] Similarly, when the Chinese government continued to pump investments into their own solar manufacturing companies, those products took over the world market.[27] And when OPEC shifted its policy on the number of barrels it produced, the cost of oil plummeted and the economics for electric vehicles became substantially less attractive.[28]

These sorts of trends are intolerable for any one company or set of investors, and private Internet and software companies face them as well. The financial crisis of 2008 hurt social and mobile tech just as much as it impacted cleantech[29]—but software companies tend to be much better shielded from external shocks. The Googles and Facebooks of the world can own their own destiny, business model, and product roadmaps to an extent that more infrastructure-dependent cleantech companies could only hope for.

The third problem is that the venture model is predicated on the efficient use of capital. In this context, VCs exercise efficiency when they build a portfolio of small, lean investments to explore myriad products and markets. When a few of these Series A investments pan out, only then does additional capital follow.[30] The cleantech model is, unfortunately and unavoidably, inverted: huge up-front costs

[25] KPCB, "Eugene Kleiner," n.d., bullet 6, http://www.kpcb.com/partner/eugene-kleiner.

[26] Steven Mufson, "Chinese Tariffs May Hurt U.S. Makers of Solar Cells' Raw Material," *The Washington Post*, 23 July 2013, https://www.washingtonpost.com/business/economy/chinese-tariffs-may-hurt-us-makers-of-solar-cells-raw-material/2013/07/23/01ac60a4-f3d9-11e2-aa2e-4088616498b4_story.html.

[27] Dezan Shira & Associates, "Who Loves the Sun? Investing in China's Solar Energy Market," *China Briefing*, 19 February 2016, http://www.china-briefing.com/news/2016/02/19/investing-in-chinas-solar-energy-market.html.

[28] Clifford Krauss, "Oil Prices: What to Make of the Volatility," *The New York Times*, 14 June 2017, http://www.nytimes.com/interactive/2016/business/energy-environment/oil-prices.html.

[29] Felix Richter, "Startup Funding Shows Signs of New Tech Bubble," *Statista*, 18 September 2014, https://www.statista.com/chart/2732/venture-capital-investments-in-the-us/.

[30] *CB Insights*, "Venture Capital Funnel Shows Odds of Becoming a Unicorn are Less than 1%," 29 March 2017, https://www.cbinsights.com/blog/venture-capital-funnel/.

must be met in order to even start the manufacturing or delivery of a product. In other words, deep investment must be made before the viability or profitability of the product can be tested.

Tesla is perhaps the most notable cleantech success story, but even that company went through five rounds of funding *before delivering a single product.*[31] That single success is of much less interest to venture capitalists than the fact that Fisker,[32] A123,[33] and ECOtality[34] needed on average $250 million of investment just to *try* to get their products to market in a sustainable way.

When cleantech investment is examined as a portfolio, the economics look even worse by comparison. It took only $2,000 for Facebook to deliver a product (our estimated cost of the software), or arguably around $13 million if the combined angel round and Series A are included.[35] LinkedIn was being used by thousands of people[36] after just $4.7 million of Series A funding.[37] These companies delivered significant returns and were on the path to success with so little capital that VCs were able to lose money on dozens of other companies at similar stages. With just Solyndra and Fisker in their portfolio, by contrast, KPCB's greentech group had already committed a good portion of their fund, so the Solyndra failure more deeply affected their bottom line.

What, then, is the future of cleantech investing?

Despite the tribulations of the past decade, cleantech investing is not going

[31] Michael Graham Richard, "First Production Electric Tesla Roadster Delivered," *Treehugger*, 6 February 2008, http://www.treehugger.com/cars/first-production-electric-tesla-roadster-delivered.html; Crunchbase, "Tesla: Funding Rounds," https://www.crunchbase.com/organization/tesla-motors/funding_rounds/funding_rounds_list.

[32] Deepa Seetharaman and Paul Lienert, "Special Report: Bad Karma: How Fisker Burned Through $1.4 Billion on a 'Green' Car," *Reuters*, 17 June 2013, https://www.reuters.com/article/us-autos-fisker-specialreport/special-report-bad-karma-how-fisker-burned-through-1-4-billion-on-a-green-car-idUSBRE95G02L20130617.

[33] Crunchbase, "A123 Systems," https://www.crunchbase.com/organization/a123systems#/entity; *Boston Globe*, "A123 Systems Timeline," 25 March 2014, https://www.bostonglobe.com/2014/03/25/systems-timeline/jngVIAkwxn48aWMr6SBTuI/story.html.

[34] Crunchbase, "ECOtality," https://www.crunchbase.com/organization/ecotality#/entity.

[35] J. P. Mangalindan, "Timeline: Where Facebook Got Its Funding," *Fortune*, 11 January 2011, para. 7–9, http://fortune.com/2011/01/11/timeline-where-facebook-got-its-funding/.

[36] LinkedIn, "A Brief History of LinkedIn," https://ourstory.linkedin.com/#year-2004.

[37] LinkedIn, "Sequoia Capital 'Links In' with $4.7 Million Investment; LinkedIn's Professional Networking Tool Accelerates Career Success and Secures Investment from Top-Tier Venture Firm" (news release), 12 November 2003, http://www.businesswire.com/news/home/20031112005300/en/Sequoia-Capital-Links-4.7-Million-Investment-LinkedIns.

away—and in fact it will only grow from here. Cleantech is not only the best (and perhaps only) solution to global resource consumption and environmental problems, but it also has the potential to be commercially successful at a global scale. The U.S.-based companies Tesla and SolarCity have proved this potential; each is a venture-backed company that has seen billion-dollar returns for investors and has had significant market and cultural impact.

Trends in the sector are improving. Ongoing growth in the solar industry has generated an accelerating reduction in the cost of solar.[38] Since 2013, the tech sector has been demonstrating astonishing innovation and progress in battery and energy storage development.[39] A number of software-energy hybrid companies have also done very well in recent years, such as Nest, OPower, and C3 Energy. These companies are decidedly cleantech, yet have such a heavy software component that their business models and product-iteration cycles fall within the sweet spot for Silicon Valley tech investors.

To better understand the future, we must take a critical step into the past in order to connect the dots of cleantech's history. For this publication, we interviewed well-regarded venture capital investors and founders in energy companies to understand their perspectives. Interviewees have spanned early- and late-stage investing in companies and also managed a major energy company:

- **Tom Baruch** has been a technology investor for over 40 years and invested in the solar company Solyndra.
- **Susan Preston** focused on seed-stage investing in cleantech and founded CalCEF Clean Energy Angel Fund.
- **John Woolard** has advised solar investments at VantagePoint Capital Partners, led the solar company BrightSource as CEO, and oversees Google's energy projects.

These conversations with venture capitalists and industry leaders provide vital context for readers to begin to chart the course forward for the next decade of cleantech, but they are only part of the story. The policy perspectives we also provide in this publication are equally important, especially as they interweave with investment and industry.

[38] Zachary Shahan, "13 Charts on Solar Panel Cost and Growth Trends," *Clean Technica*, 4 September 2014, http://cleantechnica.com/2014/09/04/solar-panel-cost-trends-10-charts/.

[39] Jeff St. John, "SunEdison Extends Solar-Plus-Storage Ambitions with Green Charge Networks," *Greentech Media*, 24 June 2015, http://www.greentechmedia.com/articles/read/sunedison-extends-solar-plus-storage-ambitions-with-green-charge-networks; GTM Research and Energy Storage Association (ESA), *U.S. Energy Storage Monitor Q2 2015: Executive Summary*, September 2015, http://energystorage.org/system/files/attachments/us-energy-storage-monitor-q2-2015-es-final.pdf.

Policy Perspective

In order to understand cleantech policy today, it is necessary to look back at the development of what is now termed "cleantech." Starting with the Industrial Revolution, technological progress consisted mostly of the development of machines with increasing mechanical or electromechanical advantages for an ever-expanding set of uses.[40] A hundred years of new manufacturing processes resulted in productivity being increased many times over by the 1940s. By the 1950s, though, technological progress was being pursued through three new and highly visible pathways.

- Nuclear energy became available in 1945 for weapons and electric power generation.[41]
- The transistor in 1947 resulted in huge improvements in computational ability and communication systems.[42]
- Space technology resulted in the launch of the world's first Earth-orbiting satellite in 1957.[43]

Each of these breakthroughs required innovation in individual technologies as well as the challenging integration of component systems and appropriate policy in order to be useful.

[40] Industrial Revolution Reference Library, n.d., "The Second Phase of the Industrial Revolution: 1850–1940," *Encyclopedia.com*, http://www.encyclopedia.com/history/encyclopedias-almanacs-transcripts-and-maps/second-phase-industrial-revolution-1850-1940.

[41] World Nuclear Association, "Outline History of Nuclear Energy," December 2017, http://www.world-nuclear.org/information-library/current-and-future-generation/outline-history-of-nuclear-energy.aspx.

[42] Priya Ganapati, "Dec. 23, 1947: Transistor Opens Door to Digital Future," *Wired*, 23 December 2009, https://www.wired.com/2009/12/1223shockley-bardeen-brattain-transistor/.

[43] Steve Garber, "Sputnik and the Dawn of the Space Age," *NASA*, 10 October 2007, https://history.nasa.gov/sputnik/.

During the Cold War, these three sets of technologies had widespread national and global impacts—first through the arms race and accompanying race to space, and then through the use of the resulting technology to improve industrial productivity and way of life for the industrialized world. Early fuel cell and solar cell technologies were developed to provide power to space vehicles in locations where cost was not a primary concern (at that time, the cost of solar cells used in space was 200 times the market price of electricity on Earth).[44] Computers became as important to the success of the space program as the powerful rockets that were necessary to escape the Earth's gravitational field and make space exploration possible.[45] Without the rapid development of computers and communications technologies, these rockets and space vehicles could not have been controlled and tracked.[46] By building this technology base at a blistering pace, the United States achieved President Kennedy's 1961 goal of putting humans on the moon by 1969.[47]

Although the rapid development of computer, telecommunications, and space technology was originally motivated by military objectives, it was quickly understood that these technologies could be transformative in accelerating technological progress in civilian economies. One of the most visible manifestations of this technology transfer was in energy use, from increasingly efficient electric generation (e.g., turbines, fuel cells, and ultimately solar cells) to advanced transportation technologies (jet aircraft, high speed trains, etc.).[48]

The energy technologies developed during the post-World War II period relied largely on abundant, low-cost nuclear and fossil fuel inputs. By the 1970s, though, world oil markets had become immense, and the cheapest oil came from what turned out to be unreliable sources because they were located in politically unstable regions of the world.[49] This situation led to much higher energy prices from 1973

[44] John Perlin, *From Space to Earth: The Story of Solar Electricity* (Ann Arbor: aatec Publications, 1999), 50.

[45] James E. Tomayko, "Computers in Spaceflight: The NASA Experience," *NASA*, 1 March 1988, https://history.nasa.gov/computers/Computing.html.

[46] David J. Whalen, "Communications Satellites: Making the Global Village Possible," *NASA*, https://history.nasa.gov/satcomhistory.html.

[47] NASA, "The Decision to Go to the Moon: President John F. Kennedy's May 25, 1961 Speech Before a Joint Session of Congress," 29 October 2013, https://history.nasa.gov/moondec.html.

[48] NASA Technology Transfer Program, "NASA Technologies Benefit Our Lives," *NASA*, https://spinoff.nasa.gov/Spinoff2008/tech_benefits.html.

[49] Richard C. Duncan, "The Peak of World Oil Production and the Road to the Olduvai Gorge," paper presented at the Pardee Keynote Symposia, Geological Society of America Summit 2000, Reno, Nevada, 13 November 2000, http://dieoff.org/page224.htm.

until about 1985.[50]

The cost of imported energy stimulated the development of new technologies to produce more nuclear and fossil-based energy at home, better renewable electric generation technologies, and more efficient energy use.[51] In retrospect, many energy-efficiency technologies seemed very accessible, but it was difficult to get individuals and corporations to quickly adopt them. In addition, renewable technologies were still significantly more expensive than conventional energy alternatives.[52] At that time, energy efficiency and renewable energy technologies were generally referred to as "greentech."

In the 1980s, energy security became recognized as a critical element of national security more generally.[53] The abrupt drop in world oil prices in 1985-1986, however, slowed the push toward energy-systems innovation.[54] Nevertheless, three basic public policy tools for stimulating energy technology innovation were emphasized:

1. Putting a price on the negative impacts of energy use on human health and welfare, which are not otherwise included in the prices consumers pay for coal, oil, and natural gas.
2. Continuing advanced research on clean energy technologies that would not normally be pursued by private firms or individuals because they are initially too far from market competitiveness, and because it is too difficult to protect the intellectual property developed.
3. Spreading information about—and demonstrating the operation of—new technology in order to accelerate its adoption by those who would benefit from it most.

These policy tools were simple to identify but elusive to implement.

[50] Chris Welles, "The Energy Crisis," *The New York Times*, 25 February 1973, http://www.nytimes.com/1973/02/25/archives/the-energy-crisis-the-struggle-to-monopolize-the-worlds-energy.html.

[51] Solar Star Technologies, "History of PV Solar," archived from the original on 6 December 2013, https://web.archive.org/web/20131206133548/http://solarstartechnologies.com/id69.html; Katherine Tweed, "In 2040, Fossil Fuels Still Reign," *IEEE Spectrum*, 14 November 2014, https://spectrum.ieee.org/energywise/energy/fossil-fuels/in-2040-fossil-fuels-still-reign.

[52] Union of Concerned Scientists, "Barriers to Renewable Energy Technologies," 20 December 2017, http://www.ucsusa.org/clean_energy/smart-energy-solutions/increase-renewables/barriers-to-renewable-energy.html.

[53] E. William Colglazier, Jr. and David A. Deese, "Energy and Security in the 1980s," *Annual Review of Energy* 8 (1983): 415–449, http://www.annualreviews.org/doi/abs/10.1146/annurev.eg.08.110183.002215?journalCode=energy.1.

[54] Dermot Gately, "Lessons from the 1986 Oil Price Collapse," *Brookings Papers on Economic Activity* 2 (1986): 237–284, https://www.brookings.edu/wp-content/uploads/1986/06/1986b_bpea_gately_adelman_griffin.pdf.

A new challenge emerged in the late 1980s and early 1990s: global climate change, caused largely by the combustion of fossil fuels and predicted to progressively worsen unless alternative energy systems could be developed. The Intergovernmental Panel on Climate Change (IPCC) was formed in 1988 to objectively assess climate change and its impact.[55] Research on climate change expanded, along with the development of new advanced energy efficiency and renewable energy technologies.

In recognition of the lack of significant impacts from the original "greentech" push, this new climate and sustainability oriented initiative was relabeled "cleantech." Unfortunately, implementation of efficiency technologies was still slow, and renewable technologies were too expensive (2X for wind and 10X for solar) to compete with conventional alternatives in the marketplace.[56]

The early days of the cleantech "revolution" were very promising as government researchers, venture capitalists, and entrepreneurs rose to the challenge; however, progress was slower than many had hoped.

First, although the cost gap between renewable and fossil energy production was getting smaller, it was still business-relevant.[57] Second, the financial crisis of 2008 made capital significantly harder to raise as financial markets across the country were affected.[58] Third, stimulated by technological innovation in the fossil fuel sector, the shale gas boom led to abundant, low-cost natural gas supplies by 2010 that competed strongly with renewables and other clean alternatives.[59]

What, then, is the recent state of cleantech?

The market share of renewables is now growing rapidly—but from a very small base, and the costs of renewables are continuing to fall despite the shale gas boom.[60] Behavioral research and improvements in IT, especially as augmented by

[55] See IPCC, "Reports," n.d., http://www.ipcc.ch/publications_and_data/publications_and_data_reports.shtml#1.

[56] U.S. Energy Information Administration, *Wind and Solar Data and Projections from the U.S. Energy Information Administration: Past Performance and Ongoing Enhancements*, March 2016, https://www.eia.gov/outlooks/aeo/supplement/renewable/pdf/projections.pdf.

[57] David Timmons, Jonathan M. Harris, and Brian Roach, *The Economics of Renewable Energy*, Global Development and Environment Institute, Tufts University, 2014, 17, http://www.ase.tufts.edu/gdae/education_materials/modules/RenewableEnergyEcon.pdf.

[58] Rebecca Buckman, "Clean Tech Hobbled by Financial Crisis," *Forbes*, 15 May 2009, https://www.forbes.com/2009/05/15/cleantech-venture-capital-technology-enterprise-tech-cleantech.html#5bdf73d6488a.

[59] Amy Myers Jaffe, "Shale Gas Will Rock the World," *The Wall Street Journal*, 10 May 2010, https://www.wsj.com/articles/SB10001424052702303491304575187880596301668.

[60] U.S. Energy Information Administration, "Renewable & Alternative Fuels: Data,"

the Internet, have removed some formerly persistent barriers to energy efficiency investments.[61] Government regulations have slowly moved to accommodate and encourage these new trends, and the entrepreneurial sector has been increasingly innovative in developing new business plans, reducing installation costs, and providing lower-cost financing for cleantech to both businesses and consumers.[62]

Thus, the second decade of the 21st century has brought new hope that energy technology innovation—and more rapid deployment of the fruits of that labor—could make the world's energy security and environmental challenges much easier to address. For this publication, we interviewed influential individuals who recognized this potential quite clearly and moved to implement public policies that could help facilitate the necessary major transformation of the world's energy systems.

These individuals all served in high-level policy positions prior to 2008, but their paths became more congruent in and around the inauguration of President Barack Obama in January 2009. Five of our interviewees were directly recruited to play major roles in the Obama administration:

- **John Holdren** had been on the President's Committee of Advisers on Science and Technology (PCAST) during the Clinton administration, was quickly tapped to be President Obama's science adviser.
- **Steven Chu** became Secretary of Energy.
- **Carol Browner** stepped up to coordinate White House energy and environmental policy development.
- **Cathy Zoi** became Assistant Secretary of Energy for Energy Efficiency and Renewable Energy.
- **Arun Majumdar** was recruited to be the first Director of the Advanced Research Projects Agency-Energy (ARPA-E). (Majumdar was actually recruited by Chu, who then tapped **William Perry**, Secretary of Defense during the Clinton administration, to chair the Secretary of Energy's Advisory Board.)

Two interviewees were major figures on the legislative side of policy development:

- **Jeff Bingaman** served in the U.S. Senate for 30 years and chaired the Committee on Energy and Natural Resources.
- **Artur Runge-Metzger** was the Chief Climate Negotiator for the European Commission.

https://www.eia.gov/renewable/data.php.

61 Ben Gaddy, "Software Ate Cleantech: Now What?" *Greentech Media*, 28 July 2016, https://www.greentechmedia.com/articles/read/software-ate-cleantech-now-what#gs.qsE3dD4.

62 Rob Day, "Cleantech Is Dead, Long Live Cleantech," *TechCrunch*, 18 April 2015, https://techcrunch.com/2015/04/18/cleantech-is-dead-long-live-cleantech/.

This unique set of individuals knew that the first phase of the cleantech revolution (roughly 2000–2010) had achieved some success but was inadequate to achieve the policy objectives President Obama and the international community had decided to pursue. So, aided by an infusion of Recovery Act money in the United States and new climate policies in Europe and California, they sought to implement policies that could help the private sector achieve governmental policy objectives more rapidly. In the interviews we present, they share what they learned from their efforts and where they think public energy policy, and especially energy innovation policy, will and should go next.

John Holdren

Former Assistant to the President for Science and Technology and Director of the White House Office of Science and Technology Policy

John Paul Holdren first developed his passion for what is now called cleantech while pursuing graduate studies in the late 1960s. He received a BS in aeronautical and astronautical engineering from MIT in 1965 and PhD in plasma physics from Stanford University in 1970. While at Stanford, he developed a keen interest in energy and the environment, partly through collaborations with noted conservation biologists Paul and Anne Ehrlich. He spent two years as a postdoctoral fellow at the Environmental Quality Laboratory at the California Institute of Technology, further expanding his expertise and interest in energy and environmental affairs. During that time, he and Phillip Herrera wrote *Energy: A Crisis in Power* (Sierra Club, 1972), a book about the full-fuel-cycle environmental and security impacts of electric power-generating technology. This work signaled the beginning of a lifelong concern about previously unknown or underappreciated effects of power generation, for example, instances where new housing developments were built on radioactive waste from nuclear fuel enrichment processing.

In 1973, Holdren joined the faculty at the University of California, Berkeley, where he co-founded the program in energy and resources. After more than 20 years at Berkeley, he became a professor of environmental policy and director of the program on Science, Technology, and Public Policy at Harvard's Kennedy School of Government, as well as a professor in that school's Department of Earth and Planetary Sciences.

While at Harvard, Holdren served as a member of the President's Council of Advisors on Science and Technology (PCAST) through both of President Clinton's terms (1993–2001), chairing several major studies at the President's request. Two studies in particular produced seminal reports: one on the R&D strategy of the U.S. energy sector, and one on international cooperation toward innovations in energy technology.[63]

[63] John P. Holdren and Samuel F. Baldwin, "The PCAST Energy Studies: Toward a National Consensus on Energy Research, Development, Demonstration, and Deployment Policy," *Annual*

Holdren was President Barack Obama's chief science advisor from 2009 to 2017, and in that capacity, he functioned as Director of the White House Office of Science and Technology Policy and Co-Chair of the PCAST. Throughout his entire career, Holdren has consistently and skillfully advocated strategies for developing and implementing new, more efficient, and cleaner energy technologies as crucial ingredients in the world's response to global energy and environmental challenges. As a presidential science advisor and the founder and leader of multiple energy programs, policies, and studies, John P. Holdren has had an important and unique bird's-eye view of the energy technology innovation process in the United States and abroad over the last several decades.

In general, what do you think is the government's role in promoting the transition of the world's economies away from primary reliance on fossil energy resources and toward reliance on clean energy alternatives?

How society obtains, processes, transports, and uses energy has a range of macroeconomic, environmental, and sociopolitical costs and benefits that are not captured in market transactions—unless government has taken steps to ensure their inclusion. These dimensions of energy choices are the main motivation and rationale for government's role in the energy domain in a market economy.

The public goods dimension includes the macroeconomic and social-welfare benefits of predictable and affordable prices for electricity and fuels, and reliability in their delivery. The externality dimension includes the impacts of energy development on environmental goods and services, the public-health impacts of pollutants from fuel combustion, the climate-change impacts of greenhouse gases from the energy-supply system, and the foreign-policy and defense-policy constraints and requirements arising from overdependence on energy imports from unstable regions.

The magnitude of the damages to public health, ecosystems, and the global climate from today's energy systems—which are dominated by fossil-fuel combustion using technologies with excessive emissions of both conventional pollutants and greenhouse gases—provides a particularly compelling rationale for government intervention to reduce these externalities. Due attention must of course be paid to doing so in ways that protect the public goods associated with affordability and reliability of energy supply. This is the biggest challenge facing contemporary energy policy, made all the more demanding because the resources, opportunities, and constraints related to energy differ from one region of the country to another, and because the transition to cleaner and more climate-friendly energy options must be global, not just national.

Review of Energy and the Environment 26 (2001): 391–434, https://sites.hks.harvard.edu/sed/docs/k4dev/holdren_baldwin_annualreview_2001.pdf.

In general, what kind of government policies do you think have been, or could be, most effective in promoting energy system transitions? Have they been as economically efficient as you would have liked? Have they been as equitable as you would have liked?

The two essential pillars of a domestic strategy to meet the challenge are: measures to promote increasing the efficiency with which energy is used to provide the goods and services that people want, and measures to promote a transition to a cleaner and more climate-friendly mix of energy-supply technologies.

In each case, three classes of policy measures are germane:

- Those that lower barriers that stand in the way of producers and consumers embracing those options for meeting these goals that are already economic, such as lack of information, inadequate financing, and the presence of perverse incentives;
- Those that incentivize embracing costlier options for increasing energy efficiency or cleaning up the energy supply, such as subsidies, efficiency, and emissions standards, or charges on emissions imposed via pollution taxes or cap-and-trade mechanisms; and
- Those that bring costlier or otherwise unproven options within reach through investments in research, development, and demonstration [RD&D] beyond what the private sector is willing to do on its own.

Important progress has been achieved in recent decades with the help of measures in every one of these categories, including efficiency labeling for cars and appliances, subsidies, and loan guarantees for investments in efficiency improvements and renewable energy technologies, a cap-and-trade approach to reduce sulfur emissions from electric power plants, and fuel-economy standards for light-duty vehicles.

The Obama administration, specifically, promulgated unprecedentedly aggressive fuel-economy standards for both light-duty and heavy-duty vehicles, greatly increased federal investments in RD&D on advanced energy-efficiency options and cleaner energy-supply technologies, streamlined permitting for renewable-energy development on federal lands, set new standards for residential and commercial appliances, partnered with manufacturing companies to increase energy efficiency in the industrial sector, and proposed the first-ever CO_2 emission standards for new and existing fossil-fueled power plants, among other measures.

There are, of course, constraints on energy-policy measures that can influence their efficiency, or their equity, or both. Many of those constraints reside in the realities of the political process—especially, currently, where action by Congress would be required. Thus, while economists generally agree that the most efficient way to reduce CO_2 emissions would be to tax them, and while the equity impacts of that approach could be addressed by rebating the proceeds to the population on an equal per capita basis, such an approach is not currently in the cards politically.

Specifically, what portfolio of policy instruments do you think the U.S. federal, state, and local governments should be pursuing at this point in time?

In terms of the federal government, the portfolio of measures being pursued under the Obama administration's *Secure Energy Future Blueprint*[64] and *Climate Action Plan*[65] constitutes a pretty good first approximation to what is achievable under current law and with the current Congress. Nonetheless, we are looking continuously for what additional sensible elements could be added under those constraints.

State legislatures and regulators have additional options open to them for encouraging investments in cleaner energy and increased energy efficiency. States also would have considerable flexibility, under the EPA's proposed CO_2 standards for the power sector, for how best to meet those standards consistent with each state's circumstances and opportunities.

In your view, what role could international cooperation play in promoting the clean energy transformation?

International cooperation on cleaner energy technologies makes sense for many reasons, which were set out quite concisely in the 1999 study conducted by President Clinton's Committee of Advisors on Science and Technology[66]—full disclosure, I chaired that study. The benefits include sharing costs and technical risks, making faster progress by bringing together the diverse capabilities of the partners, and propagating the best technical options for reducing energy risks and impacts that cross national boundaries. Such risks and impacts include overdependence on oil and gas from unstable or unreliable exporters, nuclear reactor accidents, globe-girdling conventional pollutants, and greenhouse gases.

The United States has robust collaborations to these ends with many international partners, including the European Union, Japan, South Korea, Russia, China, India, Mexico, and Brazil. As the world's two largest economies and two largest emitters of greenhouse gases, the United States and China have a particular

[64] The White House, *Blueprint for a Secure Energy Future* (Washington, DC, 30 March 2011), https://obamawhitehouse.archives.gov/sites/default/files/blueprint_secure_energy_future.pdf.

[65] Executive Office of the President, *The President's Climate Action Plan* (Washington, DC: The White House, June 2013), https://obamawhitehouse.archives.gov/sites/default/files/image/president27sclimateactionplan.pdf.

[66] The President's Committee of Advisors on Science and Technology, *Powerful Partnerships: The Federal Role in International Cooperation on Energy Innovation* (Washington, DC, June 1999), https://obamawhitehouse.archives.gov/sites/default/files/microsites/ostp/pcast-08021999.pdf.

responsibility and opportunity to lead the global transition toward cleaner energy sources, and it makes sense for them to do so as collaboratively as is practical.

Are current public and private institutions here and abroad capable of implementing these policies efficiently enough and rapidly enough to make a difference? What changes would be desirable or feasible?

Such policies are already making a difference, as evidenced by the impressive growth of the renewable energy supply in many parts of the world—including, very conspicuously, the United States—and, for example, by the fact that U.S. greenhouse-gas emissions in 2013 were about 9% lower than in 2005.[67] Of course, faster progress would be better. In the United States, a more constructively engaged Congress would be a great help.

If someone gave you $10-20 billion a year to help facilitate the energy-system transformation you believe the world needs over the next few decades, what would you do with it?

While $10 to 20 billion may seem like a large sum, in the context of the global energy system it is not. Global expenditures on energy are, I believe, in the range of $6 trillion per year, the replacement value of the world's energy-supply infrastructure is probably $20 trillion or more, and the global capital investment in new energy-supply infrastructure will likely be close to $1 trillion per year.[68]

The greatest leverage for an added $10-20 billion would be in R&D, where current worldwide renewable energy R&D expenditures by governments and the private sector combined are in the range of $30-35 billion.[69] Recent studies of what the U.S. federal government alone should be spending on energy-technology R&D, in light of the challenges and opportunities in this sector, have landed in the range of $10-15 billion per year, compared to actual expenditures of $3-4 billion per year. With an extra $10-20 billion per year for the whole world, I would put most of it into the research, development, and demonstration of a wide range of advanced energy-supply and efficient-end-use technologies, leveraging the amount through

[67] U.S. Environmental Protection Agency (EPA), *Inventory of U.S. Greenhouse Gas Emissions and Sinks: 1990–2013* (Washington, DC: The White House, 15 April 2015), Report EPA 430-R-15-004, 106, https://www.epa.gov/ghgemissions/inventory-us-greenhouse-gas-emissions-and-sinks-1990-2013.

[68] International Energy Agency, *WEO 2015 Special Report: Energy and Climate Change* (Paris, 15 June 2015), https://www.iea.org/publications/freepublications/publication/WEO2015SpecialReportonEnergyandClimateChange.pdf.

[69] Faith Birol, *World Energy Investment 2017* (International Energy Agency), 11 July 2017, 9, https://www.iea.org/media/publications/investment/WEI2017Launch_forWEB.pdf.

extensive public–private and international partnerships, and saving maybe a billion per year for incentive prizes for breakthrough advances.

In retrospect, how would you assess the effectiveness of the policies you have been responsible for implementing or encouraging Congress, DOE, EPA, and so on to implement? Do any of the initiatives you have been able to do (or encourage) stick out as having been especially successful or unsuccessful?

The way this question is formulated would make any answer presumptuous, not only because responsibility for "implementing or encouraging" is too widely shared for any one person, short of the President, to take credit, but also because assessing the relative effectiveness of policies launched in the Obama administration is premature at this point.

Has a surprising amount of competition between renewable energy advocates and energy efficiency advocates emerged over the last five years, as some have argued? If so, what do you think has caused this competition? Which side of the energy market do you think has the most potential for further technological improvement at this point in time?

I have not noticed such a trend. It has always been a challenge to avoid the zero-sum-game mentality in which it is assumed that one energy source's gain is another's loss, which can lead to advocates of one source making their case by disparaging the characteristics or prospects of another. But for the most part, even in the presence of this temptation, renewables advocates and efficiency advocates have seen each other as natural allies. The fact is that we have and we need an "all of the above" energy strategy, in which we strive to get the best out of every option. That means striving for cleaner coal, cheaper renewables and nuclear, more environment-friendly fracking, safer oil transport, and more efficient vehicles, buildings, and industries—all at once.

Do you believe a shale-gas boom that depresses natural gas prices is good or bad for the continued market penetration of energy efficiency and renewable energy technologies? Is it good or bad for energy system transitions? Good or bad for climate change policy?

The U.S. shale-gas boom has been a boon to the economy, largely through job creation, import reduction, and enhanced industrial competitiveness, and to CO_2 emissions through substituting out dirty coal plants in the electricity sector. But there are pitfalls that need to be avoided and risks that need to be minimized. As the late energy expert Lee Schipper used to say, not only is there no free lunch in the energy domain, but too cheap a lunch can give one indigestion.

The risks include the pollution of ground and surface water, fugitive methane emissions, and boom-town impacts on communities. A principal pitfall is the tempting assumption that high natural-gas use with technologies like today's can go on forever, as opposed to serving as a bridge to a future energy-supply system dominated by even cleaner options.

Do you think the EPA Sections 111(b) and (d) greenhouse gas (GHG) emissions rules[70] in the United States will be successful in reducing GHG emissions and stimulating cleantech innovation? How would you improve them over time?

Absolutely, I believe that the EPA's proposed emission rules will succeed in both respects—reducing the targeted emissions, and others, and stimulating innovation. That has been the case with sensible environmental regulations in the past, and there is every reason to think it will be the case again. The proposals embody, in their specifics, the flexibility to improve them over time, but experience with implementation—which is still to come—will be the guide to what can be constructively added or changed.

Are there any closing thoughts you have about the government's role in hastening the cleantech transition?

The U.S. government's role in the energy space is important but limited. Its investments in energy research and development are about half those of the private sector. Its investments in and ownership of energy infrastructure are far smaller fractions of the corresponding private-sector shares.

The government can shape the boundary conditions for energy choices through regulation and permitting. It can "nudge" those choices in desired directions through such measures as tax incentives and loan guarantees. It can invest in R&D on advanced energy technologies currently too uncertain of success or too distant in practical application to attract industry investment. And it can use the bully pulpit to educate and cajole.

But the large roles played by other actors in the energy domain—the private sector, state and local government, universities, civil-society organizations—mean that creative collaboration is going to be the key to success. Similarly, the fact that

[70] Section 111(b) and (d) refer to provisions of the U.S. Clean Air Act that propose limitations on greenhouse gas emissions from new and old electric power plants. See U.S. Environmental Protection Agency, "Carbon Pollution Standards: What EPA Is Doing," archived from original on 28 May 2015, https://web.archive.org/web/20150528004750/http://www2.epa.gov/carbon-pollution-standards/what-epa-doing.

the United States now accounts for approximately a sixth of world energy use and just a slightly larger share of electricity generation[71] underscores the importance of international partnerships, mentioned earlier, as a key to fostering the needed global transition to a cleaner and more efficient energy future.

Insights

Choice dimensions—from the public goods dimension to the externality dimension—must be captured in energy policy. The costs and benefits of energy consumption are not fully captured in market transactions unless the government takes steps to ensure their inclusion. These dimensions of energy choices are the main motivation and rationale for government's role in the energy domain in a market economy. The public goods dimension includes the macroeconomic and social-welfare benefits of predictable and affordable prices for electricity and fuels, and reliability in their delivery. The externality dimension includes the impacts of energy development on environmental goods and services, the public-health impacts of pollutants from fuel combustion, the climate-change impacts of greenhouse gases from the energy-supply system, and the foreign-policy and defense-policy constraints and requirements arising from overdependence on energy imports from unstable regions. The magnitude of the damages to public health, ecosystems, and the global climate from today's energy systems—which are dominated by fossil-fuel combustion using technologies with excessive emissions of both conventional pollutants and greenhouse gases—provides a particularly compelling rationale for government intervention to reduce these externalities. Due attention must of course be paid to doing so in ways that protect the public goods associated with the affordability and reliability of the energy supply. This is the biggest challenge facing contemporary energy policy, made all the more demanding because the resources, opportunities, and constraints related to energy differ from one region of the country to another, and because the transition to cleaner and more climate-friendly energy options must be global, not just national.

Depending on their stage of development and commercialization, cleantech technologies require different policies. There are a host of research and financial incentive policies that the government can use to stimulate the cleantech transition. It is important to use the right policy instruments for the level of development

[71] The United States' 2015 energy consumption was about 18% of the world total. U.S. Energy Information Administration, "What is the United States' Share of World Energy Consumption?" para. 1, https://www.eia.gov/tools/faqs/faq.php?id=87&t=1.

of each individual technology—from government-sponsored R&D for not-yet-proven or very expensive technologies with breakthrough cost-reduction potential, to performance targets for technologies close to market competitiveness, to financial incentives for technologies that are within striking distance of cost parity with fossil-fueled ones.

Government policies must be coordinated both within and between technology areas to avoid mistakes and duplications of effort, and to exploit synergies. Industry–government cooperation is essential for managing the large transitions now required, because each side has decided advantages in key areas. The public sector is better at identifying and implementing policies that change the incentives for energy innovation in the private sector to better reflect the costs and benefits of those policies for all citizens—energy producers, energy consumers, and third parties affected by energy production and use. The private sector is better at assessing and managing specific investments in response to the market and policy incentives it faces.

Steven Chu

12th United States Secretary of Energy

Steven Chu has had a lifelong interest in science and technology, and especially in developing new technologies to benefit society. He has been a world-class researcher in physics, an educator, a national lab research administrator, and a research administrator in the federal government. He is probably best known for winning the Nobel Prize in Physics in 1997 for his work on laser cooling, and for being U.S. Secretary of Energy in President Obama's first term, where he established the Advanced Research Projects Agency-Energy (ARPA-E). These accomplishments have made him one of the most highly regarded spokesmen for the cleantech revolution in the world today.

Chu earned bachelor's degrees in physics and mathematics from the University of Rochester in 1970 and a PhD in physics from the University of California, Berkeley in 1976. He stayed at UC Berkeley as a postdoctoral researcher for several years—but when offered an assistant professorship there, he went to Bell Labs instead. At Bell Labs until 1987, Chu worked on a laser-cooling project for which he and his colleagues won the Nobel Prize in Physics in the late 1990s. He left Bell Labs to become a professor of physics at Stanford University.

Moving from Stanford back to Berkeley in August 2004, Chu became the director of Lawrence Berkeley National Laboratory (LBNL); at a later date, he joined UC Berkeley's molecular and cell biology department as well as that institution's physics department. With him at the helm, LBNL focused its energy technology research program more sharply on solar energy, biofuels, and electricity grid modernization. Significantly, Chu convinced world-class energy technology researchers like Jay Keasling (a biochemical and biomolecular engineer) and **Arun Majumdar** (a mechanical engineer) to join his management team at LBNL.

In 2009, Chu was sworn in as Secretary of Energy and served there for just over four years. While perhaps best known for ARPA-E, during his tenure at DOE, he initiated other new programs to stimulate energy technology innovation, including Energy Frontier Research Centers, the U.S.-China Clean Energy

Research Centers (CERC), and the Clean Energy Ministerial meetings. During his tenure, the Energy Technology Innovation Hubs also began. These programs represented the U.S. government's most aggressive push into the cleantech arena to date, a push supported within and beyond the administration by a number of our other interviewees—Science Adviser **John P. Holdren**, White House Climate Policy Coordinator **Carol Browner**, ARPA-E founding director Arun Majumdar, DOE EERE Assistant Secretary **Cathy Zoi**, and former U.S. Senator **Jeff Bingaman** (D-New Mexico).

Chu returned to Stanford University as a professor of physics and molecular biology in Spring 2013. As a Nobel laureate who ran a national DOE laboratory, and then became secretary of energy in a cleantech-oriented administration under President Barack Obama, Chu has a unique perspective on how government policy can be used to accelerate energy technology innovation.

Can you share with us your views about public policy approaches that can speed up necessary technological transformations?

In terms of policy positions, the United States and China need to start moving more aggressively—just as California is moving as one of the largest economies in the world—to put a price on carbon. The price can start low, but it needs to rise steadily in the coming decades. The CEO of ExxonMobil has proposed that by 2030, the price should be $60 per ton of CO_2,[72] and $80 per ton by 2040.[73] His proposal gives industry time to find low-cost paths, and I believe it will stimulate incredible innovation. Without a clear price on carbon, there is very little economic incentive to develop technologies to capture carbon emissions.

As we watch the cost of solar PV modules plunge, the integration of solar power into the U.S. grid will increase dramatically. As this happens, we'll have a problem—a good problem, but a problem nonetheless. There are not yet enough innovative people thinking about how to integrate these energy sources at a large scale while still ensuring a reliable electrical system.[74] I believe there are at least

[72] See, for example, Carbon Disclosure Project, "Use of Internal Carbon Price by Companies as Incentive and Strategic Planning Tool" (white paper), CDP North America, December 2013, available from https://big.assets.huffingtonpost.com/22Nov2013-CDP-InternalCarbonPriceReprt.pdf.

[73] Kate Sheppard, "You'll Never Guess Which Companies Are Already Planning for a Price on Carbon," *Huffpost*, 23 January 2014, http://www.huffingtonpost.com/2013/12/05/carbon-emissions_n_4387532.html; Diane Cardwell and John Schwartz, "Exxon Emissions Costs Accounting 'May Be a Sham,' New York State Says," *The New York Times*, 2 June 2017, https://www.nytimes.com/2017/06/02/business/energy-environment/exxon-mobil-climate-change-lawsuit.html?mcubz=1.

[74] The "integration" here refers primarily to the fact that most renewable energy sources—wind and solar—are only available intermittently because, for example, the sun only shines during the daytime and the wind blows at certain times of the day. Thus, in running an entire electricity grid one has to either have non-renewable backup sources of electricity or storage (batteries).

a few decades of steady technology improvements—so-called experience curve cost reductions—where the cost of a technology decreases by some percentage for each doubling of production. Steady improvements in the cost and efficiency of solar and wind energy are going to continue. This will make the system-integration challenge even harder—lower costs lead to more deployment, which leads to bigger integration challenges.

The question is not whether this trend will continue, but how fast the transition to a low-greenhouse-gas-emission future can be completed. And the climate science suggests that the sooner this can be done, the better.

Wind is now—with the Production Tax Credit—cheaper than new natural gas electricity generation in the U.S. Midwest, if we assume the price of natural gas over the approximately 50-year lifetime of a new power plant is $4.00 per million BTUs. In less than a decade, wind generation is going to become cheaper than natural gas generation, even without the tax credit.

However, there is a huge difference between renewable-electricity generation and fossil fuel generation. Wind and solar power are intermittent sources, and as their penetration increases, there will be increased demands for back-up power generation, energy storage, and distribution and transmission systems that can maintain grid reliability while dealing with two-way energy flows and rapid changes in real-time generation. The situation becomes very different when intermittent power generation increases so that it becomes a much more significant fraction of the total power generated. The "grid integration" challenges can be made less risky and expensive through advanced technology and operating systems that might result from the application of a broader systems approach to utility planning and operations.

Meanwhile, solar prices are also coming down. There have to be some incentives to urge utility companies to start thinking of the systems approach. Quite candidly, more research in grid integration should definitely be a high priority, and working very closely with utility companies to actually deal with ever-increasing renewable-generated electricity is essential.

In a certain sense, we need to look to the future and, in the words of hockey great Wayne Gretsky, "Skate to where the puck is going to be." It is essential to ask, "How do we take advantage of these lower prices?" If you are a utility company, it is much easier to meet power requirements by continuing to use power-on-demand energy sources, but with intermittent renewable energy where the "fuel" is free, you want to maximize the use of the energy they generate.

Another thing that utility companies are worried about is solar energy generation on the roofs of buildings and homes. The utility companies are particularly worried about small-scale solar because of so-called third party financing models in which companies own, operate, and maintain the solar system they installed on homes, and the homeowners buy the electricity at a lower cost than they would from the utility suppliers.

Many of the present regulatory rules allow rooftop solar energy to be sold back to a utility company at retail energy prices, which are roughly twice as high as the price a utility distribution company pays. The utility company argues, "We are responsible for providing the backup power, and in order to survive, we must charge a high monthly fee to hook your solar-powered home to the grid." The utilities need to charge high connection fees for home solar because they bear the cost of the transmission/distribution systems so there is backup power.

The current dynamic begins to penalize the poorest parts of our society: the renters, the people who do not have the financial means or education to take advantage of inexpensive solar power. Policy "nudges" are needed to guide changes to the utility electricity distribution business model, and talented people to find the combination of policy and technical solutions to these challenges. Government support of research is also needed to develop the new, cost-effective technologies to take full advantage of lower-cost renewables. The future of renewable energy will be a mixture of utility-scale generation at the best sites with long-distance transmission—the most economical resources are not close to cities—local generation, and storage.

These are what I call policy actions, where I say we have to start planning. We have to start thinking. We have to start determining how we get the private sector to make those investments in the new infrastructure.

Again, the good news is we have a rapidly aging transmission and distribution infrastructure, so we have to replace it anyway. The bad news is, if we do not take policy action, they are just going to replace it with exactly what we had before, and that would be tragic.

I've heard that when PG&E has an outage, they don't even know where it is automatically, let alone how to fix it. They wait for people to call it in, and then they send a truck out to investigate.

If true, that is downright silly. There are methods of pinging power lines to help localize the outages. In the implementation of the Recovery Act,[75] we distributed thousands of phasors[76] for free. The data from those devices could trace faults automatically over larger distribution areas, and can be used, in principle, to anticipate when the AC phase is slipping and there is a danger of a local blackout spreading. However, the computer systems that the utility companies are using are not capable of handling the massive amounts of data in real time, so this projection is not currently possible. Some parts of our system are so old that some operators need to get on a phone and ask, "Hey, can you send me more power?"

This is a place where we have to encourage the firms responsible for electricity

[75] The American Recovery and Reinvestment Act of 2009 or ARRA was one of the most important elements of the Obama administration's policy repose to the great recession of 2008–2009.

[76] A phasor measurement unit (PMU) is a device that can measure instantaneous voltage, current, and frequency at a specific location in the power grid.

distribution—those firms used to be vertically integrated utility companies. The utilities are no longer vertically integrated and there is a lot of merchant power generation sold through next-day energy markets.

We have to help the electric utility companies innovate as they make this transition to becoming more market-based businesses by providing small economic incentives to stimulate the transition to more modern transmission and distribution systems. Smart meters, for example, would allow them to precisely determine the scope of an outage in residential and commercial areas.

When Stanford University started an ARPA-E smart meter project, part of the deal was that the university could use the utilities' smart meter data and make it publicly available. Amazingly, the utilities raised concern whenever someone asked them for even a little bit of extra data processing work. They would say, "We can't go to our IT guys—they have bad equipment, and we're over their data processing capacity already."

They have antiquated systems. They also sometimes say smart meter data is of "commercial interest" and detailed knowledge of energy use are proprietary secrets, which is sad.

As an example, I had to call the electric utility executives back into the Department of Energy. We had given them all of these phasors for free and they weren't sharing the information. I said, "Come on, guys. We gave you this hardware to help you avoid blackouts and enhance grid stability, and I am hearing that you don't want to share the information among each other for fear of losing some hypothetical competitive advantage." The purpose of the meeting was to shame them into moving in the right direction, and it worked.

It is also a sad reality that if you are really a technical hotshot, your first choice may entail looking to Google or Facebook rather than going to work for a utility company.

Is the trick to get those major IT companies to do some of this work?

Yes, if possible. If the leadership of those companies would say, "We have to move in this direction for the sake of sustainability, for the sake of the planet," some of their hotshots would be working to help save the planet.

There are a bunch of people in the private sector now who are nibbling at the edges of going in this direction, but they are also wary of wading into a heavily regulated industry. The distribution companies that are sitting in the heart of these challenges are realizing there is great potential in energy efficiency and more efficient use of assets, but they have not traditionally been the originators of new technology, unlike AT&T and its Bell Labs when it was a telecommunications monopoly. Research universities, including Stanford, have only recently realized the importance of research in transmission, distribution, and power electronics in general.

One more thing I would comment on is that sometimes things in energy transition strategy discussions bifurcate into either policy or technology directions. I would love to see more of the policy specialists become deeply knowledgeable about the technology. Non-technical policy wonks mostly understand the technology from a rearview mirror perspective—not what it can be 5 years, 10 years, 15 years from now. An appreciation of what is technically and economically possible over this time scale is important because infrastructure transformations take decades.

A rear-mirror view of existing technologies may allow you to make a reasonable extrapolation based on those technologies, but you will not see the new technologies being invented in research labs that could be game-changing. When I went to Washington, I discovered that most of the policy laws were not technology laws. They were written by social scientists, lawyers, lobbyists, and people with other backgrounds, but seldom by technical nerds who can envision the future.

Is it not your job to prove the learning curves wrong or the new product diffusion curves wrong?

Yes. For example, we had the Energy Information Agency within the Department of Energy. The people I worked with in that agency were mostly economists. They would make conservative technology projections. I told the rest of the DOE that it was our job to prove them wrong, and to develop new technologies that will move down the learning curve faster. In the EIA's projections of renewable energy, some of their projections of renewables' future costs were higher than what had already been achieved.

At Stanford, one of us—John Weyant—and his research group looked at the algorithms in the models for choosing new technologies. The resulting paper wound up being about how far off the models had been about the rate of introduction of wind and solar energy, because previous models used a market-share methodology that was always biased against resources with a low initial market share. They assumed that you cannot grow the adoption of any new technology for 20% per year for too long, which is right—unless you start at a very low level, and then you can. And the United States did.

The amount of electricity generated by wind approximately more than doubled and solar power increased tenfold during the time I was Secretary of Energy[77];

[77] Energy Information Administration, "Electric Power Monthly," 27 February 2017, Table 1.1A,

this generation was assisted by our loans program. If you call that a technologies innovation position, that is a very strong one. New technologies reshape all the choices and the economics of those choices.

In retrospect, how do you feel about those innovative institutions such as ARPA-E, the Frontier Centers, the Energy Innovation Hubs, and so on? Would you do those programs differently now?

For ARPA-E, I would follow the same path. The overall structure followed from the National Research Council committee report, *Rising Above the Gathering Storm.*[78] I was on that committee and was active in developing the concept of ARPA-E that originated from that report.

After the report was released, I was asked by the National Academies to represent the committee in a Congressional hearing of the House Science, Space, and Technology Committee that was considering the authorization of ARPA-E. In this case, an "authorization" bill was needed to allow the DOE to establish ARPA-E as a new funding program. An "appropriations" bill provides the actual funds. Then I was sent to the Department of Energy to try to convince then-Secretary Sam Bodman to support the report's recommendation to establish an Advanced Research Projects Agency-Energy [ARPA-E] within the Department of Energy.

The one major thing to note was that our committee proposed that ARPA-E be put under the Office of Science, because we thought that was the best functional fit within the Department of Energy.

The career people in the Office of Science were firmly against the ARPA-E proposal—they were afraid that if you started to mix the applied research goal of ARPA-E with the more basic but mission-oriented research supported by the Office of Science, the applied programs would divert funds from more long-range and fundamental research. I tried to convince them that applied research in new energy technologies was analogous to basic scientists who worked on applied programs during WWII—notably the Manhattan Project and radar—and it was a moment in history where research scientists, working side-by-side with great engineers, could dramatically advance technology. I failed to convince the Office of Science, and the House Science Committee authorized ARPA-E to report directly to the Secretary of Energy.

In terms of the internal structure and operation, it was literally modeled after my experiences at Bell Labs. At Bell Labs, the management was lean and comprised some of the very best scientists, who made the funding decisions. The organization of ARPA-E was designed to be very flat, and like Bell Labs, open and frank

https://www.eia.gov/electricity/monthly/epm_table_grapher.php?t=epmt_1_01_a.

[78] National Research Council, *Rising Above the Gathering Storm: Energizing and Employing America for a Brighter Economic Future,* (Washington, DC: The National Academies Press, 2007), http://www.nap.edu/download.php?record_id=11463.

discussions between program managers were encouraged. As the first director of ARPA-E, Arun Majumdar understood the advantages of this kind of organizational structure. He was not employed at Bell Labs, but he understood the culture we wanted to establish, and he completely and absolutely delivered all that I hoped for—and more.

What about the other programs—would you change anything?

The execution of the vision of the Energy Hubs was not uniformly successful. I wanted to recreate a concentration of scientists, similar to a miniature version of MIT's "Rad Lab" for developing radar or Los Alamos for the atomic bomb. I was concerned that research professors would band together as a "paper collaboration," where they would construct a collaborative proposal, but once funded, a lot of the money would be divided among the separate research groups defined by individual principal investigators and their groups, working quasi-independently. The intent was to have groups of people with different skills and knowledge working across disciplines and frequently communicating with a much wider cross-section of researchers.

When designing the Hubs, I wanted them to be funded for 10 years so that they could attract exceptional talent to come together to tackle hard problems. However, the Hubs were promised funding in 5-year increments, and if they were not living up to our expectations, they would not be renewed. The mangers would be held accountable—not by micromanaging them, but by letting them know that renewal was not automatic.

If there are management issues, it is important to act early on those. Acting quickly is especially important in this case, because the only reason we are funding these collaboratives doing advanced energy technology development is because there is a belief that the sum of the parts has to be significantly more than the whole.

My vision of the Hubs was for the leaders to have reasonably wide latitude in setting their research directions within the broad scope of their original charters. As an example, I was thinking that the Fuels from Sunlight Hub could investigate direct sunlight to hydrogen conversion, sunlight to electricity to electro-fuels, or hybrid systems of microbial and materials science approaches, depending on developments. I did not want DOE people in DC narrowly defining directions or being overly prescriptive, lest they close off promising developments that take researchers a bit off the initial proposal's track.

There was at least one Hub I felt was not on target in terms of what we expected in terms of synergies among groups of previously independent researchers collaborating with each other for the first time.

That is part of what made ARPA-E so good. My sense is that there were technical discussions quarterly. The program managers actually knew what was going on down in the field. They would really know what was going on. When I was a department head at Bell Labs, I knew a lot of the technical details of what the half-dozen principal investigators were doing. It is your business as a manager to get a reasonable sense of what is going on.

Again, one of the things that I thought was good about ARPA-E was that there were times when we stopped their funding with no hard feelings. It did not mean you could not apply for anything else; it was meant to be bold. We were accepting failures—we were expecting failures, but it was okay.

What do you think is the role of international cooperation in the technology transformation that is needed?

International cooperation is incredibly important for certain problems like carbon capture, or energy efficiency, or appliance incentives, or things where there are not really markets. It turns out that when we set up the Clean Energy Ministerial Dialogues, that group included roughly two dozen countries that account for 80% of the carbon emissions in the world and 90% of the world's economy. The idea we had was to get these two dozen countries to cooperate, which was unlike the UN, where you try to get 192 nations to agree to something.

It's not about trying to shift money from one country to another. It's about identifying the best practices in these countries so that others can choose voluntarily what they are interested in.

It turned out that South Africa had no idea how well minimum appliance efficiency standards could help a country achieve its objectives. They were being told by manufacturers from other countries, "No, it's going to cost you more," but actually those manufacturers were using South Africa as a dumping ground, selling old technologies there that cost more than new ones. Once you point out what's going on to the recipient of this antiquated technology, their eyes open up.

The existence of energy subsidies in many developing countries also defeats energy efficiency measures since they decrease the incentive to save money by saving energy. Clean energy at an international level through the energy ministers had a direct bearing on the governments and opened their eyes.

China has efficiency standards that are higher than what they are allowed to sell to other countries. The fact that less efficient appliances are good enough to sell as exports but not good enough to sell at home should be used as an incentive for importing countries to raise their standards.

International cooperation is where we actually see many illustrations of benchmarking: how to build more efficient buildings, how to plan more efficient cities and city areas. That is something that our architects need to learn about; even architects who claim they are "green" may not be as green as they claim. With the use of modern design tools and some simple design principles for new buildings, many large, low-cost opportunities to save on energy use can be realized.

What about shale gas? Good? Bad? Handy?

I am in favor of shale gas, but only if it is developed responsibly. There have been examples where exploration and production shortcuts were taken. The fear of regulation is so strong that the first impulse seems to be to deny that mistakes have been made. Known "best practices" in the oil and gas industry could avoid numerous mishaps.

The proper handling of backfill water is one example. Sometimes produced water brings up heavy metals and other contaminants that are not treatable in conventional sewage plants. Re-injection of produced water into deep underground formations well below freshwater aquifers is possible, as long as the amount does not induce significant seismic activity. Proper cement bond logging as one drills through freshwater tables is another example of a good drilling practice. There is no need to use diesel fuel as the "slick water" lubricant since there are better, nontoxic fluids that are of comparable cost.

All those things are fixable and enforceable if an oversight system is set up properly. Now, in the end, natural gas is a fossil fuel that still produces greenhouse gas emissions, but natural gas can help us transition away from coal more quickly. It will help China if they can do it.

Do you think China can make the transition away from massive use of coal?

They are determined to cap and then decrease their use of coal, and have formally announced their intentions before the Paris climate conference in December 2015. Air pollution has become a major problem in a number of major cities like Beijing, and the Chinese government is worried about their increasing CO_2 emissions. They are closing down their old inefficient coal power plants rapidly, and replacing them with new gigawatt-scale coal plants with modern scrubbers of SOx and NOx.[79] No country in the world is capturing carbon dioxide except in a few demonstration experiments. We have yet to prove that we have a cost effective technology that captures most carbon dioxide emissions. I believe we can increase costs by a quarter or a third to capture and sequester the CO_2, but doubling energy costs, even given the serious risks of climate change, will not be acceptable. Realistically, if a CO_2 capture system doubles the cost of producing electricity, I doubt it will be used by China or the United States.

[79] SOx (sulfur oxides) and NOx (nitrogen oxides) contribute to acid rain and ground-level ozone (smog), and are precursors to fine-particulate matter that can penetrate the lungs and cause numerous adverse health effects. See Emanuele Massetti et al., *Environmental Quality and the U.S. Power Sector: Air Quality, Water Quality, Land Use and Environmental Justice*, Oak Ridge National Laboratory, 4 January 2017, 3, https://energy.gov/sites/prod/files/2017/01/f34/Environment%20Baseline%20Vol.%202--Environmental%20Quality%20and%20the%20U.S.%20Power%20Sector--Air%20Quality%2C%20Water%20Quality%2C%20Land%20Use%2C%20and%20Environmental%20Justice.pdf.

We do need more research on carbon capture. On the sequestration part—we are doing experiments in the United States on the sequestration part amounting to megatons per year, but there are issues there as well. There are seismic effects that we are concerned about—there is some evidence that CO_2 injection increases seismic activity, but it is not conclusive at present.

We are going to have to capture carbon dioxide from all sources, not only from coal and natural gas, but also cement and coke and other large emitters. An attractive technological solution would be if we could devise a means to re-cycle the CO_2 by combining it with hydrogen to produce, ideally, a liquid hydrocarbon fuel. The hydrogen should be produced from CO_2-free energy sources such as renewable energy.

Are government-supported demonstration plants necessary to get these technologies in use anytime soon?

Yes, but it depends on what you are trying to demonstrate. If you can project how much the technology costs, and it is too expensive for China and India, I do not think we really want to spend billions of dollars demonstrating a technology that is unlikely to become cost effective at scale.

Going back to what we can share with other countries, the United States can share all these things. The United States should be doing this because there are growing markets worldwide. China and India are huge markets that are growing fast.

What India is building today—we should help them build energy-efficient buildings, industries and infrastructures, and build them economically. But they are just following all the things that we have historically done and not taking advantage of the newer, cleaner technologies. It will be a huge missed opportunity and very sad for them to follow our mistakes as well.

If all we say is, "Don't do what I did—do what I tell you to do," it may appear to developing countries that we want to slow their development. We need to convince developing countries that we are not trying to sell them anything except what is best for their own self-interest. Meanwhile, the United States should be the technology leader because we are a wealthy country, and we are still the most innovative country in the world.

If somebody gave you $10-20 billion a year to help to work against climate change, what would you do with it?

That is a good question. I would use the money to do create a Bell Labs for energy research; the laboratory would attract some of the best minds to create technologies that will change the clean energy landscape. I would also create an organization that spurs the development of self-sustaining and profitable business, showing that investments in sustainable energy can be very profitable and weaned from government support.

Would this be somewhere between ARPA-E and something VC-based?

ARPA-E is a funding agency for research and development. The technology basis for the private-sector investors such as VCs was usually initiated by research done in universities, national laboratories, and industrial laboratories willing to invest in long-term research. There is no research laboratory equivalent of telecom's Bell Labs devoted to the science and technology of sustainable energy and climate mitigation. The Bell Labs for energy would be an R&D laboratory, creating an environment in which the very best people could pursue science that would devise new approaches to energy

Unlike at a university, where a beginning scientist may be given $1 million in start-up funds, Bell Labs did not give you a whole pile of money. They gave you enough funds to give you a good start. We also knew that if you had good ideas and needed significant additional funding, say for a $100,000 piece of equipment, the decision whether to fund the idea was made—usually within days—by the department head.

Bell Labs was a very idea-rich environment. You were not allowed to assemble a big group. At most, the members of the technical staff have a technician and a postdoc. If you are not-so-good, you can have one or the other. That structure forced people to talk and work with other principal investigators; it is an ideal research environment for scientists in the first 5 to 10 years of their research careers.

In a university, typically you finish your postdoc, and then all of a sudden you are expected to become a research administrator. You hire students, but in order to nurture and train them, you also have to let them make mistakes. Students and postdocs learn by doing, and now the assistant professor is trying to learn how to be an effective administrator "by doing" as well. How professors learn to motivate and train young people is a large component of success in academia.

When I was at Berkeley I was a student, a graduate student, a postdoc, and after my eighth year, I was appointed as an assistant professor. I took the job, but since I had already spent eight years there, the Physics Department gave me the choice of either starting my own group immediately or taking a leave of absence without pay for the first two years to go to another place. I accepted the job and decided to spend two years at Bell Labs. In the end, I stayed at Bell Labs for nine years, and had the opportunity to get better and better at doing research with my own hands.

The ability to work in that special environment was a huge advantage. Many people who have spent time at Bell Labs have become leaders in university research settings as well.

What is needed to create such a thing for cleantech?

You would need at least $5 billion as an endowment after initial start-up costs. The institution would need to give the management and the researchers considerable freedom. An endowment would allow all the work to be internally funded, just as all the work at Bell Labs was internally funded. I would want to put the research center next to a major research university, and—ideally—also next to a national lab. As an example, Stanford land at the SLAC [Stanford Linear Accelerator][80] site and Arun Majumdar, the first director of ARPA-E, would be great choices.

Insights

The government can help spur early-stage technology development. There is a big role for government in sponsoring potential pre-commercial cleantech ideas, with ARPA-E and other new institutions showing great promise in this regard. A strong next step would be a Bell Labs-type institution focused on advanced energy technology development. Such an organization would provide an environment in which some of the best and brightest young minds could work together and innovate, without the large administrative and teaching burden that now falls on junior faculty members.

Manage carefully any government support for large cleantech demonstration plants. There are a few technologies for which a well-conceived and -operated demonstration plant could have a significant societal payoff. For these technologies, there is a reasonable chance that both technical feasibility and economic competitiveness can be demonstrated at commercial scale through a demonstration plant. Carbon capture and sequestration technologies are good candidates for this demonstration, since they are desirable everywhere, especially in China and India over the next few decades.

Pricing GHG emissions could stimulate technology development. A price (even a small price, initially) on greenhouse gas emissions can stimulate the development and diffusion of new, clean energy technologies throughout the marketplace. Greenhouse gas emission reductions anywhere reduce climate change impacts everywhere, providing a strong extra motivation for such efforts.

[80] The Stanford Linear Accelerator Center (SLAC) was renamed SLAC National Accelerator Lab in 2008 at the direction of the U.S. Department of Energy. The research program broadened from its original focus on high-energy physics to include strong photon science and particle astrophysics programs. See SLAC National Accelerator Laboratory, "Stanford Linear Accelerator Center renamed SLAC National Accelerator Laboratory" (news release), 15 October 2008, http://home.slac.stanford.edu/pressreleases/2008/20081015.htm.

Arun Majumdar

Former Director, ARPA-E, United States Department of Energy

Arun Majumdar has had a lifelong passion for working on and managing advanced energy technology development, especially if the innovations produced can help satisfy pressing national and global needs. He has worked toward this objective as a researcher and high-level research manager in business, government, and academia, and has even been referred to as an "energy innovation rockstar."[81]

Majumdar holds a BTech from Indian Institute of Technology Delhi and also MS and PhD degrees from the University of California, Berkeley—all in mechanical engineering. After graduation, he held faculty positions at Arizona State University; the University of California, Santa Barbara; and the University of California, Berkeley. While a professor at UC Berkeley, Majumdar started working on advanced energy technology development at Lawrence Berkeley National Laboratory (LBNL) as a scientist; subsequently, he was asked to lead LBNL's Environmental Energy Technologies Division, which has a long history of innovation in energy technology assessment, as well as in markets and policy.

For a brief time, he also served as the associate lab director under then-director **Steven Chu**. This position gave him responsibilities spanning LBNL, providing an across-the-board look at innovations in energy supply, energy demand, and energy storage technologies. At that point, Chu began formulating a number of ideas about ways to stimulate more rapid energy technology innovation that they would later bring to the U.S. Department of Energy (DOE). One idea became a response to a congressional request for recommendations that would strengthen U.S. leadership in science and technology: the Advanced Research Projects Agency-Energy (ARPA-E), a research-sponsoring government organization modeled on the military's Defense Advanced Research Projects Agency (DARPA).

After Chu became the Secretary of Energy, he recruited Majumdar to serve as

[81] Katie Fahrenbacher, "Energy Innovation Rockstar, Former ARPA-E Director, to Join Google.org," *GIGAOM* , 17 December 2012, https://gigaom.com/2012/12/17/energy-innovation-rockstar-former-arpa-e-director-to-join-google-org/.

the founding director of ARPA-E, which he did from 2009 until 2012, reporting directly to Secretary Chu. The administration budgeted $400 million to fund ARPA-E and its first projects. Since its inception, the agency has funded over 400 energy technology projects, many of which have shown early success, including a 1 megawatt silicon carbide transistor the size of a fingernail, microbes that use hydrogen and CO_2 to make liquid transportation fuel, and a near-isothermal energy storage system using compressed air.[82]

After leaving DOE in December 2012, Majumdar became Google's Vice President of Energy, looking at technology innovations, in addition to charting a course for Google Energy at a broader level. He continued to keep a foot in Google even as he transitioned to Stanford University full-time in August 2014 as a professor of mechanical engineering and a senior fellow at the Precourt Institute for Energy. Through his work in academia, government, and industry, Majumdar has a wide range of high-level experience in the development and diffusion processes of new energy technology.

Based on your ARPA-E experience, what do you think the government's role is in advancing the transition of the world's economies from relying primarily on fossil energy resources to relying largely on clean energy?

One role is certainly to fund research, the research that is too risky for the business community or the energy industry to initiate—not just to take part in the research, but to initiate it. I am talking about research on ideas that are far-reaching and, if successful, would be disruptive. Industry by itself is not going to generate disruptive ideas all the time, although sometimes it does. The role of government is to invest in research: some of it will be basic research, and some of it will be translating basic scientific concepts into a technological idea that someone has but no one has demonstrated yet. Funding risky research is a very important function of the government, because no one else will do it.

Some people in the policy development community think the VC community predominantly funds proof-of-concept[83] research; in other words, it is funding demonstrations of technology that may or may not be close to market competitiveness. Can you comment on that?

What I have been talking about is pre-venture. The venture community predominantly looks at technological ideas that have been developed elsewhere, and then

[82] ARPA-E, "ARPA-E History," n.d., http://arpa-e.energy.gov/?q=arpa-e-site-page/arpa-e-history.

[83] Proof-of-concept funding by the VC community can be successful in certain tech areas like web applications, but in the parts of cleantech where major thermodynamic, materials, and systems integration breakthroughs are needed, the capital requirements and risk involved (sometimes measured as "capital efficiency") cannot be justified.

tries to see how products developed from those ideas can be marketable.

There is a difference between technology and product. *Products* are something you can sell and make a business out of, but *technology* happens before the product. For example, someone has an idea of putting things together and making a unique functionality of energy storage at a much higher capacity. It is just an idea in that person's mind, but they have to be able to demonstrate the technology. That is a science project—there is a lot of science and engineering that goes into proving a technology can work pre-venture formation.

Sometimes VCs do fund unproven conceptual ideas,[84] but that is rare. VCs normally come in when someone has developed a first prototype of something and they have an idea about productizing it to create a business. There are exceptions to this rule, but I am just talking about the norm.

The government's role is to go pre-VC and develop the basic understanding of how the science and engineering works; in the example I gave, this would be how energy can be transformed for storage or conversion to some other form. The government certainly has to play a role in investing in research into this kind of basic technology concept.

Do you also feel the government has a central role in setting public policies to influence energy consumers?

The other side of the technology transition is energy innovations, in the market or in the ways in which people consume energy. There are two ways to influence that: a price signal or a regulatory signal. If there is no price signal or regulatory signal, nothing changes.

Appliance standards are a great example of setting a regulatory signal without a price signal. Without regulatory standards, I do not think refrigerators would have become as efficient, or cars as efficient, as they are today in the United States. The regulatory framework becomes a driver, and it actually promotes innovation. The data have shown that with a regulatory framework and the right kind of regulatory signal, consumer prices come down because you create competition and innovation in the appliance market.[85] The government can play a circular role: when the price signal is overly expensive for the consumer, government can provide a regulatory signal to the manufacturers of energy-using equipment. If consumers pay less to acquire the same level of energy service of heating and cooling, mobility, and whatnot, everyone can be made better off.

Then the question is, "Can the government provide a financial incentive to

[84] Such investment is probably more common in web-based applications and related IT areas than in cleantech.

[85] National Research Council, *Energy Research at DOE: Was It Worth It? Energy Efficiency and Fossil Energy Research 1978 to 2000* (Washington, DC: National Academy Press, 2001), http://www.nap.edu/read/10165/chapter/1.

create more competition?" The role of the government is really to create competition and provide more options to the consumer—to the people. Sometimes you may have to subsidize something for a finite amount of time so that it introduces scale and the cost comes down. If the focus is on creating competition and one thing is a little more expensive than the other options, and if there is a possible road map—a technological road map—to bring the cost down, then you can use incentives to create the competition.

Sometimes the government needs to provide subsidies to enable scale so that the costs come down. Otherwise, you have a chicken-and-egg problem where the cost does not come down because there is no scale, and there is no scale because the costs have not yet come down. The government can play a role, *for a finite amount of time*, to provide that scale, break the chicken-and-egg cycle, and jump-start the positive feedback loops that might never get initiated without it.

Do you see a government role in ensuring public health and safety in the introduction of new technologies and regulation of existing ones?

The right policy approach depends on the sector. In the nuclear area, for example, the influence of the government is much greater than in power electronics,[86] because nuclear has nonproliferation issues and waste issues, and laws are created in response to those. The government has a larger role to play in those kinds of things; traditionally, that has always been the case.

Given that I spent several months during my time at DOE to help stop the BP oil spill,[87] I can say there is certainly a government role in maintaining public safety regarding energy-related things. You can call it a regulatory framework, whether it is safety standards or the like. The government can definitely play a role in ensuring that the citizens of the United States are safe from environmental issues.

Then, of course, people have endorsed Secretary Shultz's proposal for a revenue-neutral carbon tax.[88] There are many policies like that one where there is a role for the government in leveling the playing field, by bringing in externalities to shift the economy and creating markets for cleaner energy.

There are multiple roles for the government to fill. There is a research and development role, a regulatory role, a financial policy role, and a role to level the playing field as well.

[86] *Power electronics* guide how electrical power is controlled and converted—between different currents, voltage levels, and frequencies—throughout the grid.

[87] Majumdar refers here to the explosion and subsequent oil spill at the Deepwater Horizon off-shore oil platform on April 20, 2010 in the Gulf of Mexico. For more background, see Richard Pallardy, "Deepwater Horizon Oil Spill of 2010," *Encyclopaedia Britannica*, http://www.britannica.com/EBchecked/topic/1698988/Deepwater-Horizon-oil-spill-of-2010.

[88] A revenue-neutral carbon tax adds environmental costs for the energy source to its price, but unlike other taxes, does not increase government revenues, and therefore does not affect the overall economy. See glossary for additional details.

A worldwide energy-system transformation is by necessity an international project, but some people are frustrated by the lack of desire to engage in technology transfers to major developing countries like China, for fear of them being used in a competitive manner against firms based in the United States. Where do you stand on that? How much international cooperation should there be? In what part of the new technology lifecycle can such cooperation be useful, and how can that cooperation be managed while respecting intellectual property rights and so on?

There is a place for cooperation and there is a place for competition. If you are talking about the business world, I think that businesses have to figure out whether to compete or cooperate. I am not sure that the government can or should be controlling it.

However, there is always a question about where the governments can cooperate—and frankly, if you are talking about global warming and climate change, those are global problems. They are not issues for a particular country; they involve all of the major countries—the United States, China, India and the European Union as a whole. They have to cooperate in order to come up with an understanding of how to curb emissions, and so on. Cooperation is an area in which we could do better.

What people always forget is that industries such as batteries, solar, and wind typically involve global supply chains, and different countries can specialize in different parts of the supply chain. For example, in the solar industry, China dominates in manufacturing and bringing down the cost. That is great, but the technology required to do that often comes from the United States, Germany, or some other country. Today, new technology development is happening rapidly in the United States.

There is a mutual benefit when there is competition involving different companies, particularly since many of these companies are global. They specialize in certain parts of the supply chain so that there is competition within a certain sector, but there is also cooperation. We cannot say that the United States is competing with China. The businesses are competing, not the governments—but the businesses are cooperating, too.

We need to be aware of the context of how these businesses run, and from the government's point of view, we also need make sure that international laws are followed in terms of trade. That's where many international discussions should happen so as to enable business to innovate and deploy new technologies as quickly as possible.

Often this international trade issue is framed as countries competing: one country competing with another country. But in reality, it is actually businesses that are creating and adding to the economy, and they are both competing and cooperating. We just want to make sure that the playing field is level for all of that to happen across countries. That is really important.

In retrospect, how would you assess the effectiveness of the ARPA-E funding that you were responsible for? In particular, what do you view as big successes that could be replicated by your successors, or areas to avoid?

That depends on how you define success. If you define success as "truly changed the economy," or "made a major contribution to economic growth somewhere," I do not think the impact is there yet. The story is still being told, and it is going to take time to play out. We should not raise our expectations and hopes unrealistically. However, if you measure success by how much funding it has crowded-in from the private sector for some of the projects that are promising, the multiplier is quite significant, i.e., it has attracted more capital from the private sector than the public-sector investment.

You became the Vice President for Energy at Google. Based on those experiences, do you have a different perspective than you did at ARPA-E, in terms of major industries being able to continue developing key technology?

The role of ARPA-E is to create new techno-economic learning curves for technologies that have a shot at being disruptive and transformative in the future. To give you an example, if you look at the transformation of horse carriages to cars, people could have made horse carriages better and better—they could have made better wheels, more efficient carriages, or bred bigger horses, but they could and did do something completely different in the creation of the automobile. Incremental change is one way to transform things. ARPA-E's job in the equivalent sense is to create the automobile—to create cars. Initially, a lot of the new technologies will fail—as they did for alternatives to the horse and buggy—but eventually, the technology will be disruptive and transformative for the energy sector. That's the role of ARPA-E in energy.

Many ideas will fail, but some of them will succeed—we just don't know which ones. It is very important for ARPA-E to sponsor risky, high-potential payoff research because companies, industries and businesses can take the successful ideas and technologies, create products, and sell those products. At places like Google there is a lot of very nice engineering that underlies what Google offers, but at the end of the day Google has to make products—whether they are software products, hardware products, or combinations of the two—and sell them to consumers or businesses.

The role of ARPA-E is to create that technological foundation, but businesses have to pick that technology up and then scale it, because governments and universities are not going to do it. Business will have to do it. That is the role business will have to play as a complement to ARPA-E's role.

Over the past several years, there has been competition between the demand side and the supply side of the energy technology business. Do you think there is competition? Is it good, or is it bad?

Actually, there are three sides to it. There is a demand side, a supply side, and the third is what lies between: the infrastructure. All of them are important. At Google, we spent a lot of time and effort in understanding electrical power because we need it. We need to understand that for our internal data centers.

Coming back to the demand side, it is absolutely critical. The cheapest power plant that you can build is the one that you do not have to build. So, it is important that we have the demand side, but the demand side by itself is not going to be enough. You have to have some supply—otherwise, what is the demand for? It is important to look at it as a system, and not just as the demand *or* supply side. The system includes the infrastructure in between the two.

What is going on right now in the transmission, distribution, and storage parts of the energy system is disruptive. The demand and supply sides of the big generators are also being disrupted, because natural gas is so cheap—but when the price of electricity from rooftop solar goes below that of retail electricity without subsidies, that is a disruption. You could call that a supply-side disruption. If you have local generation and storage, the storage is also getting cheaper, so then the questions really are, "What happens to the grid? What happens to utilities? Does the institutional structure of the entire electricity industry change?" Furthermore, the cost of wind and solar is coming down to the point they will be cheaper than fossil-based resources. However, they are intermittent in nature and the grid was never designed, built, or operated for such generation sources. The point is that for the first time in over a hundred years, the paradigm of centralized base load generation, and generation always tracking the load, is being questioned. And this paradigm is likely to shift.

How does a system respond to these shifting paradigms? That is a very intriguing question. I don't have the answer, but we are seeing the discussion about opening up the electricity industry structure away from its current highly centralized and regulated paradigm toward a more decentralized and flexible organization. In the next few decades, we are likely to see some significant changes.

Let's talk about aging infrastructure. Initially there were some thoughts that everything had to be immediately replaced, but IT people tell me that you can marry a lot of the existing electric infrastructure with existing Internet capabilities. Without giving away any Google secrets, do you think there is promise in pursuing that strategy?

First of all, whether we like it or not, all our infrastructure is not being upgraded at the same time. We have about a trillion dollars of assets that are not going to be simultaneously upgraded. We do need upgrades, though, so the questions really are, "What do you upgrade first? How do you plan it out? What is the road map for upgrading?"

The IT part is probably cheap because networking is cheap, and the data being collected and transmitted is cheap. There will be a level of automation that will come in as well. A lot of people are trying to figure out what to do with the data—how to use it to optimize the grid and bring down the cost even more—and that is being sorted out now. There are start-ups trying to do this.

At some point, the hardware has to be in a good place. To give you an example, the average age of substation transformers we have on the grid is 42 years.[89] That is the average in the United States. The life expectancy of these transformers is 40 years. The average transformer on line today has minus-2 years of life left already, and the backlog for acquiring transformers to replace the ones that are wearing out is 6 to 24 months.[90] That's a really big deal.

We need to think through whether we want to replace old transformers with traditional transformers, or whether we can think cleverly about this and see if we can move straight to smarter electronics, power electronics. One of the things that we started at ARPA-E is the major investment into power electronics—to go up in voltage, go down, convert DC to AC, AC to DC. New power electronics that enable the flow of electricity to be controlled very precisely, as it is done at a much smaller scale in microprocessors, could be extremely helpful, not just in modernizing our grid but to actually make it much more reliable than it is today. That is an area that will flourish. I hope it flourishes. That is the kind of thing we will hopefully see in our infrastructure upgrades, but it's not going to happen overnight.

Are there any other messages you would like to communicate to the readers of this publication about energy technology innovation?

The one thing I would say is to first take a historical perspective. We are at the bleeding edge of a 250-year-old industrial revolution. That industrial revolution is all about energy, and it has really been based upon fossil energy. Now we are seeing that the roots of this immense success, this exponential growth in our GDP and

[89] The Department of Energy's 2012 calculations put this figure at 40 years; subsequent news reports indicate that it has not improved since then. See U.S. Department of Energy, *Large Power Transformers and the U.S. Electric Grid*, June 2012, 20, https://energy.gov/sites/prod/files/Large%20Power%20Transformer%20Study%20-%20June%202012_0.pdf.

[90] Ibid., 8.

prosperity—at the very roots of this success is a potential threat, and that's climate change.

The question that people often ask is, "Should we keep the fossil fuels down to save the environment, or should we burn all the fossil fuels, because who cares about the environment?" I would say that is a completely false choice, because it doesn't take into account the science and engineering leading to innovations and technology that can completely change the ball game. Therefore, the two options need not be mutually exclusive; rather, energy and environment can be mutually inclusive.

What we really need to create are the technological foundations as well as the required policies to bring clean energy technologies to market—to create what I call "A New Industrialized Revolution."[91] Since the first Industrial Revolution was all based on fossil fuel, we definitely need the energy, but can we create a new Industrialized Revolution that is sustainable in the long run? That is going to take the next several decades, and probably the century, to create. It could be disruptive: it does not need to be, but it will probably be disruptive. It will require innovation in technology, finance, and business. And it actually offers a tremendous opportunity. So, to your readers, I say, I hope they take up this opportunity and really change the world.

Insights

Government should initiate and fund risky research. Majumdar focused on research with the potential to create major disruptions in the long-term, but which cannot attract sufficient private-sector investment because the risk of failure is high or it may take too long to pay off for normal commercial or venture investments. If the development succeeds, however, it will be close enough to commercial readiness that it can become part of the technological base upon which the venture community can draw.

Government can create market pull. Putting a price on greenhouse gas emissions or enacting emissions standards can help pull new clean technology into the marketplace, once those technologies get close to becoming competitive in the existing market. The logic is if a tax is put on the "bad" energy technology—in this case, GHG emissions—consumers will demand and supply less of it and more of the cleaner energy. This kind of incentive would make all cleantech options more

[91] This reference is a variation on the themes that noted futurist Jeremy Rifkin developed in his book. See Jeremy Rifkin, *The Third Industrial Revolution: How Lateral Power Is Transforming Energy, the Economy, and the World* (New York: Palgrave MacMillan, 2011).

profitable with respect to dirtier CO_2-producing energy technologies. However, a GHG tax would be particularly useful and essential for carbon capture and sequestration technologies. Adding CCS capability increases the base cost of oil, gas, coal, or biofuel technology, and therefore the dirtier version of these energy technologies will always be cheaper, unless there is an emissions-related premium such as a GHG tax.

The future is about the changing relationship among supply, demand, and infrastructure. It is important to look at new energy and supply technologies not separately, but rather in conjunction with each other and with connecting infrastructure. Continual reductions in the cost of IT have made pursuing smart grid technology an attractive opportunity to innovate. IT improvements and behavioral research are enabling a paradigm shift from the classical "adjust supply to meet demand" balancing to a more holistic "adjust both supply and demand simultaneously" to meet consumer needs most efficiently. With high penetration of renewable energy sources on the grid, some observers believe that the classical paradigm can even be reversed to an "adjust demand to use available supplies" balancing paradigm. In addition to IT technology and behavioral research, implementing the new paradigm will likely require both new business models and accompanying institutional innovations.

Cathy Zoi

Former Assistant Secretary for Energy Efficiency and Renewable Energy (EERE), United States Department of Energy

Cathy Zoi has spent the last 30 years working tirelessly and skillfully to stimulate the development and market introduction of energy-efficient and renewable-energy technologies in government, industry, and academia in the United States and abroad. In her career, Zoi has teamed up with former Vice President Al Gore and was a key player on the energy policy team that Secretary of Energy **Steven Chu** brought into the Obama administration to accelerate the transition to cleaner energy sources and lower energy use. As Chu describes in his interview in this publication, the members of his energy policy team felt that in order to limit the worst impacts of climate change and ensure the security of energy supplies, a far more rapid transition to clean energy is essential.

Zoi earned a BS in geology from Duke University in 1983 and an MS in resource system engineering from Dartmouth College in 1985. During the early 1990s, she pioneered the Energy Star Program[92] as a young program manager at the U.S. Environmental Protection Agency. In 1993, she moved to the Clinton White House as Chief of Staff for Environmental Policy, working closely with Vice President Al Gore.

From 1995 to 2006, Zoi worked in high-level positions in industry and government in Australia before returning to the United States in 2007 to reunite with Gore. At his request, she became CEO of the Alliance for Climate Protection, an organization established with Gore's Nobel Peace Prize award money and other donations, including proceeds from his award-winning book and documentary film *An Inconvenient Truth.*

Zoi continued her work on U.S. energy policy from 2009 through 2011, first as an influential Assistant Secretary of Energy for the Office of Energy Efficiency

[92] Energy Star was the first example of an appliance labeling program wherein the most efficient refrigerators, ovens, clothes dryers, etc. receive an "Energy Star" label, making consumers more likely to buy them and more willing to pay a higher price because of the item's higher efficiency. See glossary for additional details.

and Renewable Energy (EERE), and then as acting Under Secretary of Energy for Secretary Chu. During her time at the DOE, the organization distributed $16.8 billion in funding to accelerate cleantech deployment and energy efficiency retrofits under the American Recovery and Reinvestment Act (ARRA).[93]

After leaving the DOE, Zoi returned to Silicon Valley to become a partner at Silver Lake Kraftwerk and then later served as Chief Strategy Officer of C3 Energy. She was also a consulting professor at Stanford University from mid-2012 until mid-2015, teaching classes in cleantech innovation before accepting the CEO position at SunEdison Frontier Power in April 2015. Zoi's extensive career has enabled her to see energy policy and clean energy technology development from public, private, and academic perspectives at home and abroad over several decades.

What do you think should be the government's role should be in promoting the transition of the world economy away from a primary reliance on traditional energy resources to relying on clean energy alternatives? Should it be to correct environmental externalities? To reduce market failures in the diffusion of new technologies into the marketplace, like consumers' lack of good information, limited decision-making ability, and inadequate financing?

It is all of the above and then some. The government's role is to see the challenges that are emerging on the horizon and then navigate a way for society and the economy to get through those, while creating something good on the other side of the challenges.

That means everything from early-stage investments all the way through to commercialization assistance. All of those things in your list are appropriate for some government involvement and can be done well. They can also be done badly, but if they are done well—and we have demonstrated historically that we can do them well—they are critical.

One thing I would say about market failure is that, as Professor Mariana Mazzucato at the University of Sussex argues,[94] we are all too sensitive about correcting for market failures. In some ways governments have a chance to identify opportunities to create new wealth and seed innovation. That is not necessarily correcting for a market failure per se; rather, it is identifying an opportunity to help an economy grow while also meeting public policy objectives.

I would like to broaden the canvas on which you are painting the picture of the role of the government. I think about this a lot. I have overseen giant research

[93] The American Recovery and Reinvestment Act (ARRA) was designed to stimulate consumer and business spending by boosting U.S. government spending in order to pull the U.S. economy out of the global recession that began in 2008.

[94] Mariana Mazzucato, *The Entrepreneurial State: Debunking Public vs. Private Sector Myths*, Anthem Other Canon Economics Series (London: Anthem Press, 2013).

programs through the Technology Readiness Level [TRL][95] lens. At the early TRLs of a new potential technology are investments by government—government R&D in partnership with universities, and a bit later, government partnering with companies. All of those R&D investments have incredibly high-leverage, though again, only if they are done well. Everything has to be done well.

When I was at the Department of Energy, we would have competitions for the R&D money that was to be distributed. Labs, universities, and young companies compete for funds, and the subject matter experts in the government—and scores of carefully selected peer reviewers from outside of government—evaluated all of their proposals to fund some darned good research.

Project-specific grants and broad-based market incentives can provide very strong leverage when a technology gets past the pilot stage or the demo stage. These incentives include advanced manufacturing tax credits, investment tax credits, accelerated depreciation accounting methods, or anything that has to do with the tax code. In this sector, rebates that come from utilities are also accelerants—anything that is a financial sweetener to get a technology into the marketplace more quickly. At a later stage when a technology reaches a certain level of maturity, setting targets, standards, and codes provides a regulatory underpinning that supports continued market penetration and continuous improvement in a technology category. Consumer information and marketing programs such as labels and rating tools are also important and complement mandatory appliance standards, building codes, or renewable portfolio standards. These tools are designed to alert consumers to the existence of energy technologies that have benefits they might not otherwise be aware of. In the policy vernacular, we say that mandatory standards provide the floor, and voluntary labeling programs point consumers toward the ceiling, pushing the next stage of innovation.

I would want to make sure that performance targets, penalties, and a price on carbon appear in that list of tools. I have not talked about a price on carbon or a cap and trade system, but that fits into that incentive scheme and it could be—depending on how it is structured—a really important tool.

Standards and codes take less government money per se but require more political will. Personally, I look at the development of a new, emerging, and promising technology and I apply the tool that is going to be most effective for its stage of development in the ideal world. We have seen a similar approach—for instance, in the early days of wind in Denmark, the Danish national government was actually funding R&D while the local governments were creating manufacturing incentives to lay the foundation for creation of a big industry. The Danish government even provided a certification and a testing lab for wind. They were very supportive of the whole process.

[95] Used by NASA, the DoD, and the DoE, Technology Readiness Levels range from TRL 1 (Basic principles observed and reported) through TRL 9 (Actual system proven through successful mission operations), and provide a consistent yardstick to measure technologies and projects. See glossary for additional details.

In the United States in those days, we had a poorly implemented Investment Tax Credit and we did not have significant and continued R&D support for the wind equipment manufacturing industry. The model Denmark had, which combined different tools at different stages that sometimes overlapped with the government's direct support, was very effective at enabling Denmark to be a $15 billion world leader in wind, even though it is a tiny country.

To what extent have your programs, or others you've known about, been economically efficient and fair? How much weight would you give to those criteria in deciding what to do on the road ahead?

I am not exactly sure what you mean by "economically efficient" in the context of looking at innovation from idea stage to full-scale market transformation. One of the things that we do is measure return on government investment, and there have been macro studies on that. Colin McCormick, one of my former colleagues from DOE, has been assembling the published wisdom on this.[96]

For example, when one measures return on investment from disparate government-funded research into the seven major features of an iPhone, the literature shows that such funding has produced a giant return on investment for the world. For that reason, you can say it was good research money and good commercialization money that went into it, from the government's perspective.

Whenever the DOE issued an appliance standard, we would do a very thorough cost-benefit assessment about what it was going to cost for companies to comply with a new standard. We also considered what the benefits to society would be if the standard were in place, and the cost of carbon emissions that the public would be likely to agree to as part of these calculations. There was a government-wide review to settle on the official social cost of carbon. Everything that the DOE issued had positive benefits compared to the cost, and we would sometimes look at distributional benefits and make sure that we were not favoring just one company or one region or one demographic.

How do you respond to critics of government policies in the energy technology space? People say that either the technology or the strategies constitute corporate welfare because the government is doing things for corporations that they should do themselves.

The corporate welfare thing, I actually do not understand that beast. At the very earliest stages of a new technology's development, the government often funds very risky R&D almost completely on its own, without industry cost sharing, through

[96] See, for example, the articles in the special issue of *Innovations* 4, no. 4 (Fall 2009), "Energy for Change: Creating Climate Solutions."

the national labs they fund and university grants. At this stage in their development, the technologies are pre-commercial and, therefore, the R&D required is generally not supported by the private sector, but successful R&D at this stage can stimulate follow-on R&D in the private sector as the technology approaches market competiveness.

The reason governments do joint R&D with companies at a slightly later stage is because companies are really effective partners in getting things to market; they tend to move more quickly than a lab. So, I would argue that it is not corporate welfare. The reason that the government is involved is because there is deemed to be a public policy benefit, and the government is investing so the public benefit can be realized sooner.

That is what I would say about the corporate welfare argument. For anybody who does not like those particular things, they have the opportunity to vote the people who made those investment decisions out of office, and that is why we have a democracy.

When it comes to the appliance standards,[97] the process is so thorough and has created so many economic and environmental benefits, I am hard-pressed to find validity in the complaints—other than perhaps that the thoroughness slows things down. Every single appliance standard has to go through an exhaustive, comprehensive cost-benefit assessment, including public comments and review periods. Consumers save billions of dollars. Furthermore, every time there is a new appliance standard, appliance manufacturers figure out how to add more bells and whistles onto a technology so there is greater choice for the customer. In my experience, an appliance standard has never led to fewer features and functions, be it on a dryer, washer, or a fridge.

The claim that government investment necessarily equals corporate welfare is a silly argument. The one thing I know about engineers, and I am married to one, as you know, is that they love the challenge of having to optimize and improve. All the engineers I have ever talked to, managed or collaborated with—whether it was the engineers at Apple or IBM when we were first talking about Energy Star, the engineer designing appliances at Whirlpool, or the GE guys working on heat pumps[98]—they all say, "Oh great, you want us to make it even better. We love that!" That is what engineers do.

What role can international cooperation play in promoting the clean energy transformation? There seems to be at times an unusual concern about letting

[97] For a comprehensive overview of the process used to set appliance efficiency standards in the United States, see U.S. Department of Energy, Office of Energy Efficiency and Renewable Energy, "Appliance and Equipment Standards Program," n.d., http://energy.gov/eere/buildings/appliance-and-equipment-standards-program.

[98] A *heat pump* heats a building by taking advantage of the differential between inside and outside temperatures, and its operation can be reversed to air condition the building in the summer. See glossary for additional details.

the "other guys" get the new technologies and be more successful with them than we are here in the United States. On the other hand, if the whole world works together—and in particular with different consumer groups and different political and technical infrastructures—in principle, at least, that could be leverage to provide better solutions for all.

Seven trillion dollars changes hands every year in the energy economy.[99] Where there is a lot of money changing hands, there is a lot of money to be made. And with globalization and global markets now, there are going to be many businesses that say, "I am not giving my IP to the other guy." That's the "I do not want to cooperate" side of the argument.

On the positive side, particularly at the early stages for things that are very risky and potentially highly disruptive, there is a definitely a role for international collaboration, and even collaboration among companies. I was told the story of the International Technology Roadmap for Semiconductors [ITRS],[100] which is a good example.

ITRS started off years ago as a national semiconductor collaborative [known as SEMATECH, for Semiconductor Manufacturing Technology] in the United States, and then it went international. Apparently, the companies agreed to this road map—they would join and invest in some of the very early-stage technologies, and then share the benefits. Once the individual technology matures and they can figure out how to commercialize it, they are probably going to beat each other up in the marketplace—but in the early stage, why should each guy do it separately? I was pleased to hear that story, because in my experience I have seen that it can be tricky to get companies to agree to cooperate in this way.

What my former colleagues are doing with the SunShot Initiative is similar to that. The initiative is a national collaboration on some of the big solar challenges and there is no good reason to not take that international as well.

If you can use the ITRS as a model and carve out what is intellectual property and what is not, it would seem to me that we could get something out of this collaboration. For instance, carbon capture and usage [CCU] technologies are emerging ideas, and it would be really neat if we could get those to market more quickly through some collaborative research.

Are today's public and private institutions—here and abroad—capable of implementing policies efficiently and rapidly enough to accomplish an

[99] As of 2016, the global energy consumption market is approximately $8.5 trillion. See *Energy Consumption Global Industry Guide 2017*, MarketLine, June 2017. Available from https://store.marketline.com/report/mlig1722-06--energy-consumption-global-industry-guide-2017/.

[100] ITRS is the 15-year assessment of the semiconductor industry's future technology requirements, shared worldwide among manufacturers, universities, and national labs.

energy-system transition to low carbon energy sources and more efficient use of that energy to slow or stop climate change in time? What changes would be desirable from your point of view?

Certainly, institutions are capable of implementing and overseeing R&D effectively, of overseeing regulations effectively, of administering loan programs effectively. All of that stuff can be done effectively—or it can be done ineffectively.

ARPA-E is a great example. Since it is a relatively new organization, it is very agile and light on its feet; it runs itself just like a well-oiled research machine. There is a lot that can be learned from how ARPA-E and DARPA run their programs. They are great models for how government can stimulate research to serve national needs.

Then, on the investing in the highly disruptive side of things, we disseminated up to $30 billion of energy investment funds with the Recovery and Reinvestment Act of 2009. There was literally no audit that showed waste, fraud, and abuse because we set up systems to monitor that really well.

One of the things governments all around the world ought to be able to take advantage of is the digital revolution—being able to monitor pollution remotely with new systems that are capable of having people file reports online. That sort of updated electronic record-keeping is what government tends to be behind on.

I was just reading about the government of Estonia, when it became independent from Russia. They basically started from scratch and created a world-class government that operates extremely efficiently. The developed world and the more mature economies can certainly learn something from them.

We absolutely are capable of it. For me, one of the prerequisites is just modernizing the institutions themselves so that the government is able to take advantage of everything that business is taking advantage of.

If someone gave you $20-30 billion a year over the next few decades to help facilitate the energy system transformation you believe the world needs, what would you do with it? Or is that not enough money?

The single most important thing we need to do is have some sort of price signal on bad things like carbon emissions, so that people and firms actually have a *disincentive* to do the bad things and an *incentive* to do the good things like reducing energy demand or relying more on renewable energy technologies.

Yes, money is important, but many policymakers, researchers, elected officials, and business leaders conflate money and policy. It is not the same thing. R&D is absolutely a part of the equation and we need it, but we need a price on carbon just as much.

We need those market-pull incentives, and again, I would put a price on carbon as a market pull on that side of the equation. That alone takes more political will than dollars per se, so I would advocate that if I did not have the $20 billion, let a part of that money be collected through a price on carbon.

That would be the most useful thing for hastening the transition, because you talk to business people now and they say, "Look, we think something is coming and we do not really know what." And they quietly say, "But if you just get going and put the pricing out there, we will adjust our investments accordingly."

The pace is important, then?

We have the technologies, we just have to organize them to deploy them more quickly. You know this from the IPCC[101] work—it is the pace now. People are saying, "Well, you know batteries." The battery progress is going to happen, but we need to hasten the transition. We need to hasten the transition to 40% renewables, so we need to get all that happening faster. We need a New Deal for this energy transition: that is going to be a little bit of policy and little bit of dough [money].

Do you think the extreme energy-efficiency and renewable advocates have, at times, been their own worst enemies by over-hyping a good thing and creating unrealistic expectations about what can be achieved and by when?

No. Now that I have spent a few years here in Silicon Valley being an investor, everybody hypes their own thing. There is nobody here who does not hype their own thing.

What I love is the unbridled optimism that comes from the entrepreneur—be it a policy entrepreneur or a technology entrepreneur. As a practitioner, I apply a discount factor, but then I also ask, "But what would it take to achieve?"

Stefan Heck and I taught a course at Stanford last winter[102] in which we asked students to first do a technology and market assessment for a new clean energy technology, and then do three growth scenarios: "business as usual" or low-growth scenario, a medium-growth, and a high-growth scenario. We then talked about what sort of government intervention and business model would be needed to achieve that high potential. We know it is technically possible, and I like to think about what would it take. That is where I sit.

So much of the optimism is justified from a technical perspective, and then we have all of these other bits and pieces that kind of wear us out. We still do not have a gigantic cookie-cutter business model for accessing the energy efficiency that is available in commercial buildings. It's there, I know that it's there, but you

[101] The Intergovernmental Panel on Climate Change (IPCC) was established by the United Nations Environment Programme and the World Meteorological Organization in 1988, to review and assess the most recent research related to climate change.

[102] ENERGY 158: Bringing New Energy Technologies to Market: Optimizing Technology Push and Market Pull, Stanford University, Winter 2013-2014.

need all kinds of people in jumpsuits making upgrades building by building in a commercially systematic way. Why I am hopeful now is that we have sensors, controls and big data that can be leveraged to make a value proposition that businesses can bank on and make money from.

I don't think clean energy advocates were their own worst enemy any more than the oil industry and the coal industry are their own worst enemies. Everybody is going to hype their technology, and it takes good judgment and hard work to get the good things to grow faster.

If the utility revenue model does not get overhauled then we are not going to realize the promise here. If you want to access the great technical potential of energy efficiency, then you are going to have to put some of these other supporting mechanisms in place.

Starting about 30 years ago, the potential for energy efficiency improvements produced by bottom-up[103] studies has been a lot bigger than from top-down studies that are macro-economic and driven by energy price and quantity data. Is it possible to tap the opportunities that lie between these two types of projections? Can we provide the information necessary to affect decision-making and design policies to capture potential energy and cost savings?

Yes, but I will tell you a little story and then I will tie it back to your question. When I was 29 at the EPA, my boss at the time said, "Hey, Cathy, the fastest growing load in the commercial sector is personal computers and everybody is using them all the time. Can you design an education program to get people to turn off their computers when they aren't using them so that we can save energy on all of these computers?" And I said, "Leave it with me, let me go investigate it." And that is how Energy Star came together.

However, the insight that my young associate Brian and I had was that education programs to get people to conserve do not work very well on their own. In this case, if people thought they were going to lose the data in their computers if they switched them off, it was certainly going to be a hard slog. So, in designing Energy Star we opted instead for an automatic power-down function to save energy that *also* saved all the data in the process. That was an interesting insight, and the reason I say that Energy Star worked—right?

The reason the story is important is two-fold. One is that the cost of energy is actually quite small. To tell you another story, when I was at the DOE, my physicist colleague Henry Kelly owned a wattmeter and our head of Public Affairs, Dan, asked to borrow it. Dan had his TV on all the time so he said, "Hey, Henry, let

103 Bottom-up calculation studies gather data on energy savings for a number of specific energy efficiency improvement measures, determining the savings from each and then aggregating them to assess total energy savings.

me borrow your wattmeter so I can see how much energy I am wasting—I am keeping my TV on all day to watch the news. Maybe I won't do that anymore." So, Henry said, "Yeah, this is great. Wait until you see, wait until you see." Well, Dan plugged in the wattmeter for the little television sitting on the side of his desk. He concluded, "You know what? Keeping this TV on doesn't cost much money," and since it was actually directly related to the core of his job, he decided he would keep the TV on all the time.

The point is that when you look at all these things, even if you give people the information, you may or may not still be able to harvest that technical potential. I think that the cool thing is going to come with automated demand response technologies. Not only does the Nest thermostat make things easy for a consumer, it learns on its own and automatically adjusts by using the data it accumulates from homeowner use patterns—it remembers what consumers like so they don't have to.

Secondly, what energy engineers 30 years ago did not fully appreciate was the complexity of the "behavioral choices" made by consumers—including energy choices. With the rise of the discipline of behavioral economics, we are beginning to understand how much of what we decide in life is based on perceptions and instincts.

It is almost as if we could provide everybody with perfect, real-time information and still not be able to harvest the cost-effective energy efficiency that is in people's own self-interest. I am a big believer in the automated response that does not compromise either productivity or comfort, but just makes efficiency easy. My theory for 30 years on energy efficiency is to make it easier for people to be efficient than to not be efficient.

In retrospect, how would you assess the effect of all the funding you were responsible for during your time working for the government?

Do any of the things that were done by the Office of Energy Efficiency and Renewable Energy and other parts of DOE during that period stick out in your mind as having been especially successful or unsuccessful?

The smart grid investments were a great catalyst for getting some new utilities active in the modernizing, digitized power world. Now there is more and more data being collected on that, and it is a very cool jump-start to the transition to a much smarter electricity grid—an acceleration, if you will.

The ARPA-E investments are really nifty. Many of those, like the program we launched under SunShot, were fantastic. One program I loved was a gigantic weatherization assistance program. There were 40,000 people in America being employed to make dwellings more efficient because of this program, and they happened to be some of the neediest people.

My hope was that those 40,000 people would then move over to jobs in the private sector upgrading non-low-income residences, since the entirety of the U.S.

housing stock needs an upgrade. This is what I meant when I talked about new business models—the private sector has not quite cracked that business model. That is a bit of a disappointment to me. The work that we did that was directly funded by the government was good, but the market transformation to creating a sustainable sector in residential retrofit has not taken place yet. I wish it had. But I remain optimistic.

I love the investments that we made in solar—from new materials to streamlining local permitting. We also did some very interesting things in earlier-stage technologies like offshore wind platforms and new battery chemistries—those were fantastic.

We did some terrific R&D on solid-state lighting and supply-side emerging technologies. Some of the investments in commercialization projects have not gone that well, but when you are doing R&D, not everything works!

On the whole, it was a really good package of investments. On the carbon capture and sequestration side, we had about a dozen ground-breaking projects lined up—some were government-funded and had most had private-sector money as well. Sadly, when the Waxman-Markey bill[104] did not pass, most of those CCS pilots fell away. The private sector said, "We don't see a carbon price around the corner, therefore we aren't going to put our money to work in carbon capture."

Do you believe the shale gas boom, which depresses natural gas prices, is good or bad for energy efficiency and renewables, good or bad for the energy system transition, and good or bad for climate change?

It is good for climate change because we can see that with the data in the past couple of years, there has been very little investment in coal and much more investment in gas, so that is great.

It was a near-term depressant on renewables two years ago when investment in wind and solar went down. But I was just looking at the data and it is back up now, because we continued to move down the cost curve on solar. Solar is on a competitive basis with gas in various places now. And it will just keep getting better.

The other thing that people are realizing, after the last crazy winter in the United States, is that it is not just the price of gas—there are issues with getting gas to a freezing market because we have supply constraints. The prices spiked in New England, there was a lot of sticker shock, and a couple of competitive retailers went out of business. Low natural gas prices are probably a little bit more complicated than what originally met the eye.

[104] The Waxman–Markey bill (H.R. 2454) is better known as the American Clean Energy and Security Act (ACES) of 2009. It would have established a limit (cap) on total greenhouse gas emissions and a system for trading emissions rights from low mitigation cost to high mitigation cost firms.

Insights

Different stages of technology development call for different types of policies. Public policy measures should be tailored toward each type of technology at its existing level of development. If an energy technology is not yet technically proven, or is still several times the cost of competitive fuels but with the potential for breakthrough improvement, then it may benefit from government support of R&D to demonstrate feasibility and lower costs so that they are closer to competitor costs. If technical feasibility has been demonstrated and costs are closer—say within 50% or so—to those of the competition, then joint public–private sponsorship and government R&D subsidies can be best. As costs get even more competitive, short-term technology subsidies and technology standards can be highly effective, as they can accelerate learning about how to reduce the costs of manufacturing, installing, and using the technology. As cost parity is close to being reached, government-provided or -backed access to lower-cost financing than the incumbent technology enjoys can stimulate new technology introduction for a period of time.

Energy innovation policy requires an overall strategy. One can track the progress of each new technology through these stages of technology development and compare its performance with existing and developing technologies to put together a more comprehensive technology strategy. As the new technologies get closer to being cost-competitive in the marketplace, new business models can get them there or market incentives can get consumers to try something earlier than might otherwise be the case.

Consumer education may help taxpayers appreciate already cost-effective energy efficiency technologies. Many energy efficiency technologies are already cost-competitive but require consumer education to proliferate in the economy. In addition to education, capturing this energy efficiency potential will likely involve making consumer decisions easier, for instance, through machine learning. Government regulators should not hesitate to impose technology mandates where consumer information is bad and preferences for specific product attributes that lead to less energy consumption are weak or nonexistent.

Carol Browner

Former Administrator, Environmental Protection Agency (EPA)

Carol Browner has been a tireless advocate for the environment. She worked closely with former Vice President Al Gore in several positions and is probably best known for being the highly-influential Administrator of the Environmental Protection Agency (EPA) for the entire eight years of the Clinton Administration, making her the longest-serving person in that position since EPA was established in 1969, and also for her service in the first term of the Obama administration as director of the White House Office of Energy and Climate Change Policy.

Browner received a BA from the University of Florida in 1977 and a JD from University of Florida's Levin College of Law in 1980. Upon graduation from law school, she worked for the Florida House of Representatives. She then moved to Washington, DC, where she became a legislative assistant for Senator Lawton Chiles (D-Florida) and later legislative director for Senator Al Gore (D-Tennessee). Browner then returned to Florida in 1991, where she ran the state's Department of Environmental Regulation until 1993.

Her growing reputation as a strong and skillful environmental regulator in Florida and her close relationship with Gore were likely important reasons then-President Bill Clinton chose her to be Administrator of the Environmental Protection Agency in 1993. During this period, she reorganized the agency, introduced EPA's first partnerships with industry, strengthened U.S. air quality regulations significantly, and defended EPA's budget from attacks by congressional conservatives.

After Browner left the EPA, she joined Madeleine Albright in launching The Albright Group. She then returned to federal service as director of President Obama's White House Office of Energy and Climate Policy (2009–2011). Her position led some to refer to her as the administration's "Energy Czar" or "Climate Czar." In that capacity, Browner was part of a team of energy environment experts driving clean energy initiatives across all federal agencies, including Science Adviser **John Holdren** and Secretary of Energy **Steven Chu**.

Today, Browner resides primarily in Washington, DC, where she is senior counselor at Albright Stonebridge Group, chair of the board for the League of Conservation Voters, and a member of the Center for American Progress Board as well as other advisory and corporate boards.

What do you see as the government's role in promoting the transition of the world's economies away from a primary reliance on fossil energy resources to reliance on clean energy alternatives?

The tax credits for renewables have been hugely important. They have been effective.

Industry has suffered from Congress' inability to stay the course. They turn these credits on, turn them off, re-establish them, change them, make them retroactive, etc. All of that uncertainty has been unfortunate in that it makes cleantech investment planning very complicated, but nevertheless the existence of the subsidies has served the industry quite well. I believe the Production Tax Credit for wind and the Investment Tax Credit for solar energy could have served investors better if they had been applied more consistently and were changed with more advance notice, though.

Renewable electricity portfolio standards—specifying a minimum share of all electricity generation that must be from renewable sources—and energy efficiency standards have also been a good way to create opportunities for those sectors. We had a well-established electric utility sector that was built at a time when there were monopolies, and it was important to regulate them as monopolies.

What portfolio of policy instruments, specifically, do you think the United States federal, state, and local governments should be pursuing?

I was very supportive of what President Obama tried to do.[105] It is all the art of the possible. I would have liked to see a cap on carbon and an ambitious emissions reduction credit trading program, but those were not possible. President Obama looked at all the tools that were available to him and put them to good use. For instance, look at sector-specific work, such as requiring a certain fuel efficiency in cars and the work they are doing on power plants.[106]

[105] The White House, *Climate Change and President Obama's Action Plan* (Washington, DC, 7 June 2015), https://obamawhitehouse.archives.gov/node/279886.

[106] See U.S. Environmental Protection Agency, "Carbon Pollution Standards: What EPA Is Doing," archived from original on 28 May 2015, https://web.archive.org/web/20150528004750/http://www2.epa.gov/carbon-pollution-standards/what-epa-doing; the Clean Power Plan was a centerpiece of the Obama administration's Climate Action Plan. See Peter Foster, "Barack Obama Unveils Plan to Tackle Greenhouse Gases and Climate Change" (video), *The Telegraph*, 3 August 2015, http://www.telegraph.co.uk/news/worldnews/barackobama/11779765/Barack-Obama-to-unveil-tougher-plan-to-tackle-greenhouse-gases-and-climate-change.html.

I guess for me—being in the political world—there's the economists, and the people who get things done. The economists can tell me what is "perfect" from an economic point of view, but then there is also just the reality of "how do you get something done based on the options that are actually available?"

In your view, what role could international cooperation play in promoting the cleantech transformation?

Climate change is a big problem, and it is a global problem. The world would be best served by finding the most efficient reductions in greenhouse gas emissions, including carbon. In some instances, these emissions reduction opportunities may be in places in the United States (e.g., retiring coal-fired power plants in the Midwest or the Southeast), and in other instances, they may be in other parts of the world (e.g., building natural gas or renewable power plants instead of coal plants in China or India). I certainly think that these new clean technologies are among the most cost-effective and common-sense solutions.

Part of the problem we have as a society and as a world is that we do not look at the externalities. We do not actually think about what pollution is costing us. We simply think about the capital investment and maintenance cost of what needs to be built. It's a very narrow point of view.

What do you think about the capability of public and private institutions here and abroad for implementing the policies necessary to react to a problem like climate change?

It will take a combination of both public and private actions to respond to climate change: the best of what government has to offer, and the best of what the private sector has to offer. There are times when the government will lead and, quite frankly, there are times when industry will lead. People's clean air expectations do not change, but whom they might expect to deliver the solutions for clean air may evolve. During the George W. Bush administration, which was very retro on this stuff, you certainly saw the states stepping up and, as an example, adopting mercury standards.

I don't think there is one right answer—the best approach to environmental regulation is a combination. In a perfect world, in which the government can set really rigorous science-based standards and then allow industry flexibility to adapt to meet those standards, the best answer will emerge.

Do any of the policies that you implemented at the EPA or in the White House stick out as having been very successful or unsuccessful? If the latter, how would you do things differently?

I will give you a good example: the rule requiring reductions in emissions from diesel fuel use in cars and trucks. It had a huge public benefit because the particles now trapped used to wind up in people's lungs, often causing illness and sometimes death. We talked to truck and auto manufacturers, and they realized that they had to really think about how to reconfigure their engines. The manufacturers subsequently came to believe that this regulation helped make them first-in-the-world in these cleaner diesel technologies—and that their innovations were actually stimulated by a government regulation. Ultimately one CEO acknowledged that a tough standard drove innovation.

The second example is the CAFE[107] rules that we adopted. Those rules are part of what has made U.S. car companies globally competitive. We are making very efficient cars, and we will make more efficient cars in this country. That's big.

If someone gave you $10-20 billion a year to help facilitate the energy-system transformation you believe the world needs, what would you do with it?

I would spend the money on battery technology, and the whole world of distributed energy. How can we make better use of everything from my farm in Vermont—I have solar panels and I sell power back to the grid, but what about the top of my apartment building in DC? A lot of these options are not high tech, but require rolling things out at scale—and quickly.

The fear I have as a private citizen is that the investment we are making in natural gas as a "bridge fuel" is—if history is any guide—going to make us stuck with natural gas for a very long time. It may look attractive today, but it is not going to look very attractive 20 years from now. That it is not going to give us the emission reductions we will need by then. We would still be emitting carbon—less than with burning coal, but there is still a lot of carbon from natural gas combustion. I guess that's the bottom line. You are also building a huge infrastructure around the new natural gas supplies, and you are not investing in other things.

We saw it in coal, too. In 1990, the people who ran coal-fired power plants said, "Do not regulate us. Do not make us meet new pollution standards—we are going to go out of business if we have to do that." Twenty years later, they still say they are going to go out of business if we ask them to reduce the amount of pollution they produce.

Recently, there has been some competition between the energy efficiency and renewables communities, as if the renewable energy advocates are saying, "Forget about energy efficiency—we need renewables to replace fossil fuels.

[107] CAFE (Corporate Average Fuel Economy) Standards were originally set in 1975 to increase the fuel economy of cars and light trucks. Each auto manufacturer must meet fuel-efficiency targets on average, across all of the cars they sell.

Renewables first, efficiency second." Do you have that sense too? And if so, why do you think that exists?

I would say efficiency first—efficiency is the low-hanging fruit. We should work our way up the hierarchy from the lowest- to the highest-cost carbon mitigation options. Everyone has a different perspective. But my perspective puts a high priority on addressing the negative impacts of climate change. If I am in the renewable energy industry, my perspective is more on building a business, with climate change mitigation as only one of several important benefits to consider.

Are there any other thoughts you have that you would like to share?

One issue we did not touch on that does have a technology angle to it is adaptation and resiliency. Sadly, we now have to talk about them.[108]

I try to think of adaptation and resiliency as opposite sides of the same coin. Resiliency is how quickly you can recover from a storm, for example. And adaptation is, "How do we use nature, how do we use engineering to try and adapt to the reality we're about to encounter?" We have to think about both of them.

Insights

Policy stability and consistency matter. When subsidy and tax credit programs are switched on and off—as opposed to being phased in and out gradually—it creates an uncertain investment climate for private investors.

Pragmatic trumps theoretical for policymakers. In developing a policy architecture, it is crucial to include policies that not only have desirable conceptual properties, but also are politically and institutionally implementable. This approach may require giving up some "efficiency" in terms of lower costs, but a policy that cannot be implemented is not very useful. For example, market-based regulations may efficiently reduce GHG emissions, but if they are unfair—or perceived to be unfair—to major constituencies, they will be difficult to implement.

[108] If one is not able to reduce the impacts of climate change by reducing greenhouse gas emissions, it may be possible to reduce the damages incurred by people and businesses by helping them to adapt to the changes they experience. Examples include helping farmers irrigate, developing drought-resistant seeds, or helping low-income people buy air conditioners in areas that get hotter.

Success requires both public and private sector participation. The public sector can set clear goals and ground rules, and then leave the private sector as much flexibility as possible to meet these goals. The policy design should allow flexibility to adjust goals in response to emerging markets, business models, and technology developments. Flexible use of technology standards and subsidies can be effective in promoting public–private partnerships and accelerating the transition to cleaner energy without compromising the international competitiveness of the U.S. economy.

Jeff Bingaman

Former United States Senator and Chairman of the Senate Committee on Energy and Natural Resources

As a native of New Mexico, with its abundant fossil fuel resources and two major Energy Department National Laboratories,[109] Jesse Francis "Jeff" Bingaman developed an early interest in energy production, use, regulation, and policy. During his impressive rise through the ranks of state government to a leadership role in the U.S. Senate (D-New Mexico), he became a great believer in the role of advanced technologies in facilitating the transition to a cleaner, more sustainable energy system; as a result, he led or helped lead several major pieces of legislation over his decades of government service.

Bingaman earned a BA in government from Harvard in 1965 and JD from Stanford in 1968. After law school, he became counsel to the 1969 New Mexico Constitutional Convention. Following that, he entered private law practice in Santa Fe. In 1978, Bingaman was elected New Mexico Attorney General; in 1982, he was elected to the U.S. Senate.

He served five terms in the Senate, from 1983 to 2013, chairing the Senate Energy and Natural Resources Committee twice during that time. During his three decades in the Senate, he became a leader in the development of new energy legislation in the public interest, and energy technology innovation was a major focus of the laws he sponsored. Bingaman figured heavily in the effort to pass the Energy Policy Act of 2005 and was the lead sponsor of the Energy Independence and Security Act of 2007 (EISA).

Bingaman retired from the U.S. Senate in 2013 and returned home to New Mexico, where he remains involved in public affairs through speaking engagements. During 2013 and 2014, he served as a Distinguished Fellow at the Steyer-Taylor Center for Energy Policy and Finance at Stanford Law School, where he published a report on state-level policies for clean energy.[110] In 2015, he taught a seminar

[109] Los Alamos National Laboratory and Sandia National Laboratory.

[110] Jeff Bingaman et al., *The State Clean Energy Cookbook: A Dozen Recipes for State Action on Energy*

on the functioning of the U.S. Congress in the Honors College at the University of New Mexico. With his wealth of experiences in state and federal energy policy, Bingaman remains a leading voice on energy production and consumption.

What is the role of government in promoting the transition of world economies from a primary reliance on fossil energy resources to a reliance on clean energy alternatives?

Although we have made substantial progress in designing policies to reduce energy consumption, there is very little economic incentive at the federal level for the transition to a low carbon energy system to occur.

Within that frame, what kind of government policies do you think have been or could be most effective in promoting energy system transitions?

You need robust government policies that set targets for what needs to be accomplished. For example, if you are trying to reduce greenhouse gas emissions, it is appropriate to set standards for what is going to be permitted and then leave it up to the group that is the subject of the standards or has the standards applicable to it to find ways to meet those standards. It is best when government can stay away from choosing particular technologies to insist upon. Setting standards is what states have done with the renewable portfolio standards. They said, "We want a certain percentage of our power produced from non-emitting sources of generation," and then left it up to utilities to figure out the least expensive way to do it. There are a lot of other examples as well.

To what extent do you think energy technology standards constitute corporate welfare, and why might some people consider energy efficiency standards a violation of the principle of consumer sovereignty, especially here in the United States? How would you respond to these critics?

I have never thought there was a lot of validity to the argument that the government could not set efficiency standards on appliances, vehicles, furnaces, lighting, or anything else that is to be sold in our market. It seemed to me that it is entirely appropriate as part of the government's effort to encourage an efficient use of energy to set those standards. A side effect of that, to some extent, is that you are limiting consumer choice, but the benefit to the economy overall is substantial and

---

Efficiency and Renewable Energy (Stanford Steyer-Taylor Center for Energy Policy and Finance, 11 September, 2014), http://media.law.stanford.edu/organizations/programs-and-centers/steyer-taylor/State-Policy-Report-low-res.pdf.

outweighs any inconvenience to consumers that this might cause. Frankly, a lot of the wailing and gnashing of teeth about a loss of consumer sovereignty is more of an ideological perspective on the world than it is a real-life concern for most people.

On the research and development side, how do you respond to critics who think government sponsorship of this type of applied R&D is nothing more than corporate welfare and cite the Solyndra experience?

I think Solyndra was an unfortunate instance in which we were trying to use a loan guarantee to encourage the proliferation of new activity in solar manufacturing here in this country. Obviously with hindsight, it was a riskier proposition than many of the other things we, as a government, have tried to do with loan guarantees. It is easier to use a loan guarantee to encourage proliferation of existing technologies than it is to use the government's authority to support newly developing technologies that have to compete in the world market. Solyndra is an unfortunate example, but the overall idea that the U.S. government should try to encourage the development of a clean energy industry in this country is one I strongly support.

In your view, what role could international cooperation play in promoting the clean energy transformation?

It has to play a very substantial role if one of our key objectives in pursuing this so-called clean energy transformation is to reduce greenhouse gas emissions. Obviously, looking ahead for the next few decades, the biggest part of that problem is going to be in the emerging economies where the demand for power is growing and where emissions are also growing.

Let us say someone gave you $10-20 billion a year, over the next two decades, to help facilitate the energy system transformations you believe the world needs. What would you do with it?

I am not sure what I would do with all of it. One thing we need to do is find ways to incentivize the manufacture of low-cost, low-emission or zero-emission vehicles. Maybe some portion of that $10 to $20 billion could be used to establish a prize to accomplish that, where you'd basically say, "We'll do the same thing with low-emission or zero-emission vehicles that we did with the energy-efficient light bulb," where we set up, I think it was a $10 million prize. Philips came along, competed, and won to make an LED-based light bulb, which met some very specific criteria, and demonstrated that they were able to manufacture a substantial quantity of these and market them.

I thought that was a useful exercise, and something similar to that could be done with electric vehicles or low-emission vehicles of some kind so that we could get some competition to bring the price down. I recently got a tour of Tesla's new facility and their manufacturing operation in Fremont [California]. It is a tremendous facility with state-of-the-art manufacturing, but when you check on how many cars they are producing each day it is only 100.[111] When you check the number of vehicles manufactured and sold each day worldwide, it is 165,000. The concern I have is if we are going to be producing 165,000 vehicles per day worldwide—and much of that of course is to meet the demand in the emerging markets—we have to find a way to increase the percentage of those that are low-emission vehicles.

There was an article in *The Wall Street Journal* recently about the success of electric cars in Atlanta. Nissan sold 831 Nissan Leaf plug-in electric cars down there and they were very happy.[112] The number that struck me as particularly concerning is that they said that nationally only 0.38% of the vehicles on the road today are electric vehicles. We have a long way to go.

The prize idea is great. It sure would be nice to have a whole army of Elon Musk-type people out there competing and collaborating with each other.

The problem is, worldwide, how do we produce these vehicles in significant quantities cheaply enough that people can actually buy them? I am glad that some people are able to buy Teslas but most people are not. We have to find a way to produce low-cost, low-emitting vehicles that can be used particularly in the emerging world, but also in our country as well.

In retrospect, how would you assess the effectiveness of the energy policy programs you were personally responsible for developing—with a little bit of help from your friends? Do any of the things that you were able to do in the Senate stick out as having been especially successful or unsuccessful?

Some of the things we did were successful. They were not as successful as we would have liked, but they did move us in the right direction. One thing we did that I was

[111] That volume has increased over time. According to Tesla's 2017Q3 investor letter, the company produced 26,137 vehicles during the quarter, translating to roughly 290 cars per day. It projects that 2018 will see over 700 cars produced per day of their new Model 3 alone. *Tesla Motors – Third Quarter 2017 Shareholder Letter*, November 2017, 1, http://files.shareholder.com/downloads/ABEA-4CW8X0/5828576941x0x962149/00F6EB90-2695-44E6-8C03-7EC4E06DF840/TSLA_Update_Letter_2017-3Q.pdf.

[112] Richard Halicks, "Electric Cars Gain a Toehold in Atlanta," *Atlanta Journal Constitution*, 1 December 2014, http://www.ajc.com/news/news/local/electric-cars-gain-a-toehold-in-atlanta/njJds/; Mike Ramsey, "Atlanta's Incentives Lift Electric Car Sales," *The Wall Street Journal*, 4 June 2014, http://www.wsj.com/articles/why-electric-cars-click-for-atlanta-1401922534.

glad to see accomplished was the establishment of ARPA-E. **Arun Majumdar** did a great job of organizing that new agency and **Steven Chu** did a great job of leading the charge for it. In the long term, it can do a lot of good for the world and for our country. Some of the tax provisions we were able to put in the law have been helpful.

The Investment Tax Credit [ITC] in particular, as applied to the installation of solar panels in the country has been a good thing. We were able to pass that particular ITC provision for an eight-year period, which gave enough of a planning horizon to see a substantial increase in use of solar power in the country. It is not nearly what it needs to be but it is substantially more than it used to be.

Wind power has also benefited from the Production Tax Credit we put in place. That has not been as well designed because it was for a shorter term and we did not have an eight-year extension of that provision or an act of that provision. I think changing the law in 2007 so that we could once again start increasing CAFE standards was a very major step in the right direction. An awful lot of people deserve credit for that.

Most of those other people would give you a large share of the credit for that.

Dianne Feinstein [D-California] deserves a lot of the credit for the hard work she did on it. It came out of the Commerce committee, not out of the committee that I chaired there in the Senate, but we included it as part of the energy bill we passed through Congress and President Obama signed. That was after a 30-year period of fighting with the automobile manufacturers and the unions who were adamantly opposed to giving the Obama administration authority to raise those CAFE standards.

I am trying to recount how many times I voted to change the law so that CAFE standards could once again start to increase, before we finally got it enacted. We voted pretty much every Congress there for a long time and were defeated in each time.

You have undoubtedly seen the excitement about the shale gas boom—if it actually does depress natural gas prices over the next decade or two, is that good or bad for energy efficiency and renewable energy deployment? Is it good or bad for energy system transitions? And is it good or bad for the climate change policy efforts?

It depends on how we respond to the shale gas evolution. If we have the wit and the wisdom to continue to pursue development of renewable energy, then the shale gas revolution will have been a good thing. If, on the other hand, it lulls us into neglecting pursuit of some of these other ways to meet our energy needs then I think obviously that is a bad thing. As a general matter, the shale gas revolution that we have seen and the shale oil revolution now is a tremendous opportunity for our country. I gather it is going to spread to the rest of the world and it will be good for the rest of the world too. We need to try to keep our bearings as far as the long-term future of the planet and the long-term need of our economy—which is to continue to develop ways to meet our energy needs that are not dependent upon fossil fuels.

What is your opinion of the EPA rule under the Clean Air Act Section 111(d), which would be used to limit emissions of greenhouse gases from exiting power plants,[113] *and Section 111(b) which is being used to limit emissions from new power plants?*[114] *The proposed Section 111(d) rules would require states to limit the emissions of greenhouse gases per unit of electricity produced within their boundaries to 30% below 2005 levels by 2030.*

I guess from what I've read and the press on it, it doesn't strike me that it's that heavy a lift for many states and for many utilities. It's a good thing that they've done it. The President [Obama] deserves to be commended, the EPA deserves to be commended, and—I'm sure it will lead to litigation, controversy, and all the rest, but I think there are things underway that will allow utilities to meet these reduced emissions targets without a whole lot of heavy lifting in the process.

Insights

The appropriate type of government support varies as each technology moves through its development cycle. At all stages in the new technology development and diffusion process, the main justification for public support needs to be the prospect of capturing benefits from innovation that accrue to others. Important stakeholders include cleantech firms, other sectors in the economy, energy consumers, and citizens. With regard to citizens specifically, it is important to consider benefits for citizens who experience negative impacts (e.g., air pollution, climate change impacts, etc.) associated with energy production

[113] Megan Ceronsky and Tomás Carbonell, *Section 111(d) of the Clean Air Act: The Legal Foundation for Strong, Flexible & Cost-Effective Carbon Pollution Standards for Existing Power Plants,* Environmental Defense Fund (EDF), October 2013, revised February 2014, https://www.edf.org/sites/default/files/section-111-d-of-the-clean-air-act_the-legal-foundation-for-strong-flexible-cost-effective-carbon-pollution-standards-for-existing-power-plants.pdf.

[114] U.S. Environmental Protection Agency, "Carbon Pollution Emission Guidelines for Existing Stationary Sources: Electricity Generating Units," *Federal Register,* 18 June 2014, https://www.federalregister.gov/articles/2014/06/18/2014-13726/carbon-pollution-emission-guidelines-for-existing-stationary-sources-electric-utility-generating.

and transformation, which would not normally be considered in R&D, and also investment decision-making by individual firms or venture funds. Thus, in each case, government intervention seeks to change the conditions in which the private sector makes decisions on clean technology investments, to make them more attractive relative to dirtier options. The list of areas where government intervention can have such impacts is long, but the choice of appropriate policy instrument needs to be focused on the stage of development of the individual technology and the externalities that are preventing markets from creating the incentives that would most benefit society as a whole.

Consider prizes for the creation of new technologies with target cost and performance objectives. One specific prize could be for the development of low- or no-carbon emitting vehicles at a price that is affordable in the developing world. In addition, prizes for large improvements in performance or reductions in the cost of producing clean energy technologies (e.g., electric or hydrogen cars) could be quite effective, as already seen with refrigerators, diesel engines, and more recently, efficient lighting technologies.

Set efficiency standards. Regulatory policy instruments can be effective in achieving energy-environment objectives. Standards are especially effective when applied to large energy uses where market-based policies like a GHG tax or an emissions cap and trade system are politically difficult to implement. Efficiency standards for important energy-consuming goods such as appliances, light bulbs, and automobiles have been quite effective in reducing the energy and carbon intensities of economies around the world and, if these standards are set carefully taking into account technological and economic potentials, they can be implemented with very small negative—or even zero—impacts on consumer choices.

William Perry

19th United States Secretary of Defense

William Perry served as the 19th Secretary of Defense of the United States, from February 1994 to January 1997. He is perhaps best known as one of the principal architects of major international agreements designed to reduce the world's stock of strategic nuclear weapons.[115] For Perry, however, the road to Washington has led through Silicon Valley and back several times during his career.

Although the arms-control debate has a number of technical dimensions, Perry's previous Department of Defense position as the Director of Defense Research and Engineering (DDR&E) probably has the most relevance to his work in and views on cleantech entrepreneurship and policy. The innovative use of advanced materials and information technologies has been critical in both the smart weapons and cleantech transformations. As DDR&E, he had responsibility for weapon systems procurement and research and development; for his instrumental role in the development of stealth aircraft technology, he is often referred to as the "father of stealth."[116] Stealth was itself part of a much larger Cold War strategy at that time, which was to offset the tremendous advantage the Soviet Union had gained in military hardware, such as missiles, boats, and airplanes, with smarter systems incorporating advanced digital sensor and control technologies like GPS,

[115] Many individuals have contributed to these efforts, but the United States has played a leading role in the negotiations with a group of four senior strategists often referred to as the "gang of four": William J. Perry, former secretaries of State George P. Schultz and Henry Kissinger, and former Senator Sam Nunn (D-Georgia), who have been central in the development of U.S. and global strategies. See George P. Shultz, William J. Perry, Henry A. Kissinger, and Sam Nunn, "Next Steps in Reducing Nuclear Risks," *The Wall Street Journal*, 5 March, 2013, https://www.wsj.com/articles/SB10001424127887324338604578325912939001772.

[116] John Lancaster, "Perry is Quiet, Managerial but Unabashedly Mr. 'High-Tech,'" *The Washington Post*, 25 January 1994, https://www.washingtonpost.com/archive/politics/1994/01/25/perry-is-quiet-managerial-but-unabashedly-mr-high-tech/63394025-5f1a-4ac3-9978-bb26115ea531/?utm_term=.be9126234490.

the Internet, laser guidance, and so on. In defense circles, this became known as the "Offset Strategy."

Perry's technological roots have deep Silicon Valley connections. He holds a bachelor's and a master's degree from Stanford University, and also a PhD in mathematics from Pennsylvania State University. He worked in the defense electronics industry from 1954 to 1964, founding Electromagnetic Systems Laboratory (ESL) in Silicon Valley—a major defense electronics company—and serving as its president for over a decade.

Upon leaving the DDR&E position in 1981, Perry returned to Silicon Valley as managing director of Hambrecht & Quist (H&Q), a San Francisco investment banking firm specializing in high-tech investing. He then became a professor in the Department of Engineering Economic Systems and co-director of the Preventive Defense Project at the Center for International Security and Cooperation, both at Stanford University from 1985 until 1993. During this time, he maintained an affiliation with H&Q, primarily making investments in IT ventures—in addition to some early cleantech investments.

In 1994, Perry became the third person with a technical background to be appointed Secretary of Defense, a role he served until 1997. (The other technically trained defense secretaries were former CalTech President Harold Brown, Perry's boss as DDR&E, and Charles Wilson during the 1950s.)[117]

Perry then returned to Stanford as a senior fellow at the Hoover Institution and the Freeman Spogli Institute of International Studies. His experiences since 1997 are of direct relevance to the issues discussed in this publication, in particular his service on the Secretary of Energy Advisory Board from 2009 to 2015 for **Steven Chu** and the ARPA-E Advisory Board for **Arun Majumdar**.

Perry was awarded the Department of Defense's Distinguished Public Service Medal in both 1980 and 1981, inducted into the Silicon Valley Engineering Hall of Fame in 1996, awarded the U.S. Presidential Medal of Freedom in 1997, and named a Stanford Engineering Hero in 2013.

When not on an airplane to Washington, DC or Seoul, Perry spends his time these days working as professor emeritus at Stanford University with the Freeman Spogli Institute for International Affairs and the Schultz-Stephenson Energy Policy Task Force at the Hoover Institution. Perry has seen the technology-innovation process in the United States not only from the perspectives of Silicon Valley, broader industry, government, and academia, but also from high-level positions inside the defense, IT, and energy sectors.

Can you briefly describe the technology-oriented innovation positions you have held?

[117] Subsequently, Ashton Carter, Secretary of Defense in the Obama administration, also has a technical background and is a long-time Perry collaborator (e.g., Ashton B. Carter and William J. Perry, *Preventive Defense* [Washington: Brookings Institution Press, 1999]).

In 1964, I left GT&E Laboratories[118] to found my own company named Electromagnetic Systems Laboratory [ESL]. It was very much a high-tech company back in '64. It was designed to bring the digital revolution to the things that were in intelligence reconnaissance. It's hard to think of the digital revolution as something new and exciting, but at the time it was. So that drove our company.

After 12 years of doing that, the company had become very successful. The technology was there, companies like Intel were just starting projects out in the field, and companies like Hewlett Packard were starting to release their first compact computers, like the HP 2000. The digital revolution was an idea whose time had come, and it turned out to be very successful.

In early 1977, the new Secretary of Defense, Harold Brown, called me and explained that we were lagging behind the Russians—three-to-one in numbers of planes, tanks, and ships—and we needed technological innovation. He was looking to create something called the Offset Strategy and wanted me to be his director of defense research and engineering. Specifically, he wanted me to offset the hardware advantage of the Soviets by using the digital technology that I had been using in the private sector in military applications—if successful, this would allow us to counter the Soviet weapons-hardware advantage by creating "smarter" weapons. I didn't really want to leave my company, and this was not something I wanted to do at all. But the idea of being able to implement this strategy was too attractive, so I accepted the challenge.

At the time, I didn't know about stealth technology as one of the potential applications, but I found out about it immediately after I was sworn in. The application of the Offset Strategy was digital technology, reconnaissance systems technology applied to smart sensors, smart bombs, and so on. Stealth, in a sense, was digital technology because the design of stealth is basically done on computers. I spent the next four years developing these digital weapons technologies and bringing them to the stage where they were ready for production.

The stealth fighter, the F-117, had its first operational flight just months after I left office. We felt we had to get programs to an irreversible position by the time we left, because we only had four years to do it. Even if Carter had been re-elected, I wasn't planning to stay another four years.

Where did you go next?

In the 12-year period between my times in government, from 1981 to 1993, I was at Stanford. At Stanford, I pursued mostly policy issues concerning nuclear weapons in Russia and China, but I was also on the board of a number of high-tech companies, including a solar company and a wind company. Way back in the 1980s, solar was an idea whose time had not yet come.

[118] GT&E (General Telephone & Electric) was later named GTE. GT&E Laboratories was the firm's research center, similar to Bell Labs.

I spent part of my time at Stanford and part of it at Hambrecht & Quist, where I was doing high technology venture capital. I was perfectly content to spend the rest of my life doing those things, but I got dragged back to government again in 1993. As the Secretary of Defense, I didn't really have an opportunity to get involved in technology the way I had before, even though some brilliant technologies were coming along then.

After I finished my term as Secretary, I became involved with a number of really interesting high-tech startup companies in a diversity of fields—one in the clean energy field, one in health care, and one in computer-application education.

Besides defense, I've been involved with clean energy for the past 30 years, since my time and Hambrecht & Quist in the 1980s. All during these years I have been on the advisory boards of both the Department of Defense and the Department of Energy. I was the chairman of the Secretary of Energy's advisory board for Secretary Steven Chu. We were looking at problems like fracking, and we were particularly focused on the environmental hazards of fracking and what to do to mitigate those risks. We were also examining what could be done to advance the various clean energy programs that were in the industry, and what would be the appropriate role for government to play in advancing those programs.

As you know, Steven Chu had decided that a good thing to do was to have the government facilitate bringing good ideas into production. It worked well a few times, like DOE's SunShot Initiative for the acceleration of solar PV technology and similar programs for LED lighting and advanced battery development. It worked very badly one time, in the case of the Solyndra federal loan guarantee that got the department a black eye for having invested in manufacturing.

I still think the concept that the government has a role in bringing a good idea from the development stage into manufacturing is a good idea. If I were trying to stimulate technical progress on the manufacturing side, I would do it not by investing in a manufacturing facility directly, but by giving energy companies a long-term contract at a fixed price, which would provide the stability for the company to get capital from private sources. It's a good idea for the government to help industry make that step—not by making capital investments itself, but rather by being a guaranteed, long-term customer at a good price. Then, the companies can use those contracts to attract the capital they need from private sources.

Where do you see the government being a sole consumer? Are there other strategies, other than guaranteed price, that you think are good?

Steven Chu not only initiated the loan guarantees, but also ARPA-E and the Energy Innovation Hubs and the Energy Frontier Research Centers. So, in the cleantech space, what do you think the rationale is for those types of initiatives—and how would you personally organize them?

The government can play three roles that are quite different, but each of them is important and unique. The first, obviously, is supporting fundamental research—what the government calls 6.1-Basic Research and 6.2-Applied Research.[119] There's just very little of this being done that far up the private industry R&D system today. There's very little incentive for a company to invest in private research that other people can then take to market. The government has a unique role in investing in fundamental research—in a sense, doing what Bell Laboratories and IBM did decades ago. Universities do it, and they do a pretty good job. The Defense Department does a lot of it, and more recently the Energy Department has been doing some. There's no substitute for research, so that's a fundamental part.

The second role is taking ideas that have military application all the way to development, and they have to do that with one of the military services. The trick there is getting a service to buy into this research program, and then DARPA and the service take it forward together. That's what we did with stealth, for example. We had to sort of beat the Air Force into agreeing to do it—not because they were opposed to the idea, but because they could see it taking money away from their F-16 program. There is a limited amount of funds available for this kind of work. This is an important government role, but it applies only to ideas with military applications. Obviously, it applied to the stealth airplane, but that didn't have any civilian fall-out.

The classic example of developing a technology with military and civilian usage, which I was involved with during the time I was DDR&E, was Global Positioning System [GPS] technology, although I must say at the time we didn't appreciate the dual-use aspect of it. I was pushing the development and production of GPS precisely because I could see its military applications. The program was almost canceled, and it took a heroic effort to keep it alive during the time when I was undersecretary. It had already been canceled, in fact, and I had to turn it around. That was a case where the government played this role of helping a system that has military applications through its development and production, and, if you're lucky, there's some dual use and the civilian sector benefits from that.

This is where an organization like DARPA plays a key role, because they take the ideas that they believe have high payoff, and then—usually with the help of the secretary or undersecretary—they add on a service to go along with it, which drives it forward. As I say, most of those programs have turned out to be very successful for the [military] services, and a few of them have turned out to have dual use, like the GPS system.

[119] Similar to Technology Readiness Levels used in other agencies, the Department of Defense has its own R&D evaluation scale. The two Budget Activity categories above are followed by Advanced Technology Development; Advanced Component Development and Prototypes; System Development and Demonstration; Research, Development, Test, and Evaluation Management Support; and Operational System Development. See Undersecretary of Defense (Comptroller), *DoD Financial Management Regulation, Volume 2B: Budget Formulation and Presentation, Chapters 4–19* (DoD 7000.14-R), November 2017, http://comptroller.defense.gov/Portals/45/documents/fmr/Volume_02b.pdf.

When Steven Chu became Secretary of Energy, he wanted to have that kind of a model in energy, so he set up ARPA-E and got Arun [Majumdar] to head that. Other people have tried to emulate DARPA in the past, unsuccessfully. So, what was the difference here? Why was ARPA-E successful? There are several reasons. There was a lot of emphasis and interest in energy, so Chu sort of had the wind at his back. Secondly, he understood why DARPA had been successful in stimulating the development of GPS, Internet, and stealth technologies among others, and he set up ARPA-E to have the same chance at being successful. Finally, he got a very bright guy, a very energetic guy, to head it up. For all those reasons, ARPA-E turned out to be successful, whereas the other would-be DARPAs have not been.

The third thing that government can do—and has done on a few occasions—is search industry and apply the dual-use concept in the other direction. They can search industry for what's been done for industrial purposes and see if that could be used in the government as well. This was the exception back in the 1970s when I was the DDR&E, but it's the rule today: More things are being done in industry first and are then picked up by the military.

In-Q-Tel was set up specifically to exploit that model, which is, "Here's something that's being developed in industry—maybe it has a neat application in government, too. Let's see if we can set up a facilitating mechanism to get that transferred into the intelligence community."

The United States was fortunate to have civilian-sector spillovers—including cleantech—as benefits from the technologies developed for the Cold War and race to space. As a nation, we were innovative and a very successful leader in global problem-solving. Now that the Cold War and the Space Race have ended, are you worried about losing that sense of urgency and the flow of ideas that came from it? Is part of the push toward global sustainability inspired by that concern? Is enough being done on cleantech from the policy side to let the private sector pick up on new technologies and diffuse them?

If I look at what I see as the major advances going on in cleantech today, I don't see a very significant government hand in there. The government is in there to take advantage of private-sector advances in cleantech when available and when appropriate, but I don't see government being a leader. It is certainly not a leader in fracking, nor an absolute leader in solar energy that I can see. Whether we're talking about clean energy, or Internet communications, or massive, mass-data manipulation, the leadership is in industry—and industry, as you well know, is a global industry today.

At this point, well, you can always date the demise of large-scale government involvement in solar to the bankruptcy of Solyndra. What Steven Chu had chosen, and what he said we need, was manufacturing capability—because there's enough

R&D going on here. So, he chose one particular path to facilitate manufacturing and got badly burned on it. From about that point on, nobody in government has tried to do that sort of thing again.

You raise the very good question of where it is appropriate for the government to leverage private-sector innovations in pursuit of its own objectives. To do so, the U.S. military, space, and intelligence communities have to be very adroit on their feet, to stay abreast of what's going on and extract advances quickly. When they identify something that's not being done and that needs to be done, then they can put their own R&D on it and add their efforts.

That's not being done as well as it ought to be, partly because people who tried to do it, like Steven Chu, found that the environment is pretty unforgiving. You make one mistake and then the whole concept gets thrown out.

We were lucky back in the late 1970s, when we were trying to get this Offset Strategy going, that we didn't have any spectacular public failures. We might have. In fact, just one little anecdote I'll share with you is that the prototype for the F-117 crashed during one of its test flights. That was a problem: not only did it raise doubts about the program, it also exposed the program because it crashed out in the open somewhere. We somehow managed to keep that from the public, even from Congressional awareness. So, first we were lucky that we didn't have a really major failure. The less significant failures we did have, like that one, didn't reach the public eye, so we just charged ahead.

We were protected in that case by the very high classification of the program. That is why we had such a classification—we benefitted from the classification, but we were also lucky we didn't have any major failures.

In the cleantech, global sustainability, and energy system transformation space, what do you feel is the role of international cooperation? You've tried to work some pieces of your Preventive Defense doctrine[120] around cleantech. So, what do you think the United States government role is in this process—and should we, as a nation, be doing more in that arena from the policy side, even if it means bringing in the best and the brightest from the private sector?

I think so, for the simple reason that we in the United States benefit from other countries being clean as well. We just have one atmosphere, so the fact that China is building one coal-fired plant a week is really bad news for us in terms of CO_2 emissions. If we can find a way of helping them switch over to cleantech, should we do it? The answer is yes. *Can* we do it is much trickier, because most of the really

[120] The Preventive Defense doctrine seeks to make international conflict less likely by promoting cooperation between major global military powers—i.e., to seek to prevent armed conflict rather than only preparing to engage in military conflict. See, for example, Ashton B. Carter and William J. Perry, *Preventive Defense* (Washington: Brookings Institution Press, 1999).

significant cleantech work is being done in private industry, not the government. The advanced development work in particular is being done in the private sector, not so much for security reasons as for commercial reasons, so the people owning the IP are not too inclined to share.

What could the government do differently? For example, one thing that we could do to help lower CO_2 emissions in China is facilitate their development of fracking. Well, that's not something that the government could prevent from happening—although a lot of the early research on fracking technology was done with its support, now the government doesn't have the proprietary information about actual implementation of this technology—the oil and gas industry does. So, getting this technology transferred to China is a matter of the leading fracking companies in the United States cutting deals with the Chinese. In this case the role of the government is to not get in the way—because it could see some benefit to the United States in cutting those deals. Had it been a government-developed technology, it would be a different matter, but it in this particular case it was not.

What impact does the U.S. government have in China in terms of solar technology? Not much. Same with wind power. Even though the government had helped finance and develop the technology by helping some of these cleantech companies get started, typically it does not own the IP for the technologies as they have been implemented. It doesn't own the IP, so it can't transfer it. If transferring is a security problem, the government could prevent it. But it can't mandate it. The government can play a role in foot-dragging, but it can't really accelerate the transfer.

The answer to your original question is yes, there's a real benefit to the United States—particularly in China, to facilitate their getting away from their dependence on coal. Moving away from burning coal for electricity as soon as possible will benefit the whole world, but particularly the United States as it is increasingly being impacted by climate change.

What are your general views on fracking and the shale gas boom? When the history books are written, do you think the shale gas boom will turn out to have been a good, bad, or indifferent occurrence?

The shale gas boom is a huge benefit to the United States from many points of view, and certainly from a CO_2 emissions point of view. It's basically driving the coal-fired plants out of business. From an economic point of view and from a national security point of view, there are three really big plusses, which makes me a huge enthusiast. I'm not concerned directly about the environmental issues, because I think they're minimal. What I'm concerned about is that given the variety of companies pursuing fracking, not all of them are going to be responsible. Some of them will take shortcuts, and we may have an environmental disaster with the same kind of impact on fracking that Three Mile Island[121] had on nuclear power. That is the danger.

For that reason alone, the industry should be taking very serious steps to self-regulate—and the government, when it can, could be doing more. It's very difficult for the federal government to regulate fracking because, as you know, it's primarily a state regulation issue. Every state has a different approach to it.

That's what I worry about on fracking. There's a huge benefit to the United States, to the environment, to our national security policy, to our economy—everything. But there's an ever-present danger of some irresponsible driller having an environmental mishap, which in turn would create a PR disaster, which would then put huge pressures to slow down or stop the fracking.

Is there potential for the federal government to exert more influence on the states, say, through arguing that regulating fracking is material to interstate commerce, both in economic terms and certainly in terms of public health and safety?

I believe the answer to your question is yes, but we're not doing it. And I'd say it's unlikely we're going to be doing it, given the political environment we're in right now.

So, this puts pressure back on the private sector to be responsible. Some environmental groups, like Environmental Defense Fund (EDF), are trying to fill that space by proposing best practices for environmentally safe fracking operations. How successful have they have been?

The person who could give you more knowledge to answer that than I would be Steven Chu. I've discussed it with him in the past. He can tell you directly, but I think he concluded that the problem of federal government regulation of fracking operations is too hard. In the political environment he was working in, he couldn't figure a way to make it happen. To put it another way: as his advisor on this question, I thought it was such a difficult political problem that I didn't see fit to push him on this issue.

[121] Perry refers to a partial meltdown of a nuclear reactor at the Three Mile Island power plant, near Middletown, Pennsylvania, on March 28, 1979. Although not much radiation actually leaked out, the incident had a very strong influence on public opinion regarding the safety of nuclear power generation both in terms of radiation exposure and possible nuclear explosions; it was the most serious nuclear power generation accident in U.S. history. See U.S. Nuclear Regulatory Commission, "Backgrounder on the Three Mile Island Accident," 25 May 1979, http://www.nrc.gov/reading-rm/doc-collections/fact-sheets/3mile-isle.html.

The work you were doing when you were DDR&E was really science fiction that became reality. These technologies don't seem too far out, but something that you've done throughout your career is take technologies from one domain and transfer them to another domain in a pretty obvious way.

Yes—let me give you another example of that point. Before I went into the DDR&E job, I had been using digital technology primarily to develop space reconnaissance systems. We were in an amazing space in 1977, a very amazing space. We had developed photo systems, electronic systems, radar systems—the whole battery of reconnaissance technologies. One of my objectives as the undersecretary specifically was to take that technology and move it down to a tactical level.

We talked about smart sensors earlier as a new technology adapted for the military—getting the technology that we used in space reconnaissance programs down to the battlefield commander. We did it in two ways. First of all, by facilitating the transfer of data the very same way that we sent it to strategic intelligence consumers or down to the tactical level, and also by putting up airborne systems.

The Airborne Warning and Control System [AWACS] is one example. The Joint Surveillance and Target Attack Radar System [JSTARS] is another primary example, though. To me, JSTARS was a revolution.

AWACS monitors all air traffic over the battlefield and relays it in real time to the battlefield commander, so any battlefield commander has a display that shows him where every airplane is in his air space, which is a huge advantage, of course.

AWACS was already underway and in development when I became secretary, and so I brought that to a conclusion. But at the same time, I said, "Why don't we do the same thing for ground vehicles?" So JSTARS does for the ground-battle space what AWACS does for the air-battle space. Every vehicle on the battlefield shows up in real time: where it's located, where it's moving, how fast it's moving. It's all displayed right to the battlefield commander in real time.

When Desert Storm[122] started, JSTARS had been developed already but hadn't gone through its final acceptance tests. General Norman Schwarzkopf said, "I want that." Of course, they said, "Oh, you can't have it—it's not ready yet." He said, "I want it." The engineers who were doing the testing took it over to the battlefield in Kuwait and operated it, so JSTARS was in full operation all during Desert Storm, but it was being operated by engineers, not by military personnel. And it worked like a charm.

[122] On August 2, 1990, Iraqi military forces, under orders from dictator Saddam Hussein, invaded Kuwait. Operation Desert Storm was the U.S. military's code name for the operation that began on January 15, 1991 to repulse Iraqi troops from Kuwait in what became known as the Persian Gulf War. Evidence of the availability and use of the new high-tech weapons systems described by Secretary Perry here were quite evident in this conflict and were often featured on evening news reports. For more background see History.com, "Persian Gulf War," 2009, http://www.history.com/topics/persian-gulf-war.

Is there any other advice or commentary you want to make on the whole issue of the government's role in facilitating the transition to clean energy sustainability on a global scale?

I just would draw a contrast between today's clean energy drive and what we were doing in the late 1970s in the Offset Strategy, when we were bringing the digital age into tactical weapons systems. The contrast is—we were in charge. We knew what we wanted. We were funding the technology: we derived some benefits from industry, but by and large our programs were self-standing.

In a sense, what we were doing was relatively easy compared to what needs to be done today, because today the key technologies are not government technologies and they are not being funded by the government. They might have been funded by the government at an early stage, but not at the commercial stage.

First of all, the government has to understand what's going on in the development of the new technologies, what these technologies are being designed to do, and how they can be applied in cleantech—and then figure out how to get it transferred from the private sector to where it can be used. It's a transfer from industry to government now, from the private sector to the public sector.

It's a more difficult problem than I was faced with back in the 1970s when we were able to pay for much of the technology development with federal funds on national security grounds and then transfer the new technologies to the private sector afterwards, but it's no less important because the private sector is where most of the technology is today. It's not in government laboratories—it's in Google, in Facebook, or at Twitter. That's where we find the technology that's most important to us.

Even in the space field today, where decades ago we had a transfer of government space-imaging systems to a private company that was starting to build their own imaging systems—today, the government is using some of those private systems. The resolution they're getting is good enough to be really very interesting. More and more, what the Defense Department needs in particular, and what the government needs more generally, is technology that's already being developed and applied by industry. It needs to deal with a transfer in a different direction than what I was dealing with. When I was DDR&E, it was interesting to see that the technology we were developing for GPS had a dual use and was transferred into the civilian sector. It happened much sooner, much more quickly than I'd imagined possible. It's a different problem today and, in many ways, a harder problem.

Insights

Balance the public and private role of technology innovation and commercialization in the cleantech sector. As cleantech innovation continues to become more IT-oriented, there should be a concomitant emphasis on technology transfer from the private sector to the government sector because these technologies tend to evolve faster in the private sector. However, in some areas like solar cells and biofuels, government-sponsored energy science and advanced engineering research can help provide new materials, physics, and chemistry foundations for commercial development of new clean energy technologies by the private sector innovation.

Risk management matters when allocating government funds in cleantech. Where government funding is involved, it is a good idea to keep the amount of money at risk to manageable levels. This approach minimizes the political repercussions of the inevitable failures that will occur in response to risky but high-potential payoff R&D investments. Although these investments are risky, government oversight should be well-informed and systematic in order to avoid big mistakes. The Solyndra case serves as an example of where oversight was lacking; however, most of the SunShot Initiative R&D investments reflect well-executed oversight.

Government can help bring new technology from the development stage to the manufacturing stage. To stimulate technical progress on the manufacturing side, rather than investing in a manufacturing facility directly, the government can give the energy companies long-term contracts at a fixed price. With the government as a guaranteed, long-term customer at a good price, the companies can use those contracts to attract the capital they need from private sources.

Susan Preston

Founder, CalCEF Clean Energy Angel Fund

Susan Preston is the founding general partner of the CalCEF Clean Energy Angel Fund. In 2007, CalCEF hired her as a consultant to guide them in creating a seed-stage fund to support their clean energy efforts. The resulting Angel Fund is a for-profit limited partnership, separate from the CalCEF (California Clean Energy Fund) organization, which is a nonprofit mutual benefit corporation. CalCEF is the founding limited partner of the Angel Fund; other limited partners include CalPERS, large pension funds, wealth managers, foundations, and family offices. The Fund focuses on seed- and early-stage investments in the clean energy sector, and has invested within the United States in companies based in California, Colorado, and Tennessee. Portfolio companies include Alphabet Energy (waste heat recovery) and Boulder Ionics (ionic liquid electrolytes for advanced battery technologies).

Preston started her career as an environmental lawyer at the forestry company Weyerhaeuser. She then served in several positions in public and private companies from general counsel to CEO, including medical device therapeutic company MicroProbe. She is a patent attorney and has been a partner in three different law firms during her career. She also founded the first all-women's angel investment organization in the United States, Seraph Capital Forum.

Regarded as an expert in angel financing and angel organizations, Preston spent six years working as an entrepreneur-in-residence with the Ewing Marion Kauffman Foundation and continues to serve as a global consultant focused on angel investing and entrepreneur initiatives. She has written numerous articles, white papers, and books, including *Angel Financing for Entrepreneurs: Early-Stage Funding for Long-Term Success* (Jossey-Bass, 2007) and *Angel Investment Groups, Networks, and Funds: A Guidebook to Developing the Right Angel Organization for Your Community* (Kauffman Foundation, 2004). Preston's experiences in investing at an energy startup's earliest stages—when there is often only an idea, team, and prototype—are invaluable to understanding cleantech.

How did you become interested in cleantech investing? How did you get started?

My interest in clean energy, cleantech, and the environment goes way back to my days in college and law school. My bachelor's degree is in microbiology and microchemistry, so I have a scientific background. After law school, I was an environmental lawyer at Weyerhaeuser Company, a large global forestry company based in the state of Washington. I was lucky to work for a company that took environmental stewardship seriously and had already spent close to a billion dollars in environmental safety projects before I arrived. Through that experience, I became interested in and concerned about the environment and how we analyze climate change.

After Weyerhaeuser, I went into biotech in senior management positions. I became a general counsel at a medical device therapeutic company, then was a lawyer and a partner in three different law firms. In 1999, I started doing angel and venture capital investing. I started Seraph Capital Forum, the first all-women's angel investment group in the United States, here in Seattle.

I was then connected with the Kauffman Foundation and ended up working there for about six years, first on women entrepreneurship initiatives and then in angel investment activities. During my time there, I published two books and closely studied early-stage investment structure, including angel, seed-stage, and venture capital.

Soon thereafter, a nonprofit in San Francisco called the California Clean Energy Fund [CalCEF], approached me. They were interested in establishing an early-stage seed fund to support their clean energy efforts. I was first brought on as a consultant focused on fund structure options. They then asked me, "Why don't you come down to the Bay Area and see if you would like this?"

As a result, I moved down to the Bay Area in 2007. I wrote the offering documents and started raising capital, with the first close in February 2008. I became the general partner for the CalCEF Clean Energy Angel Fund. It has been a great opportunity to create and run an early-stage fund focused on clean energy.

You have written numerous articles and books on angel financing. What is one important thing an angel investor, as opposed to a traditional venture capital investor, needs to consider? And then, for cleantech specifically, what do angel investors need to consider?

I don't think the parameters are any different for a sound investment whether you are an angel investor or venture investor. My firm belief is that you have to understand that numerous factors will influence any company's opportunity for success—the number and extent of each risk is more challenging to quantify at the very early stages.

For cleantech in particular, the market is absolutely paramount and the regulatory environment is critical. That's where a lot of people completely missed the beat. Many VCs or angel investors who had been successful in other areas like semiconductor investments thought energy was just another cool market area. That is far from the truth. The energy sector is a very difficult area and the routes to technology adoption are fortuitous. All factors related to market, technology, and team must be considered when analyzing an investment opportunity.

In the early days of the clean energy investment rush, I think the difficulty was discounted and many people did not understand that, or were not even aware of it.

One of the biggest risks to any early-stage investor is what you don't know. You need to be able to discern what the risks are and determine the path that the company should take to either reduce or eliminate those risks. But when you don't know that a risk is out there, you can't quantify it nor can you identify the path to eliminate it.

Whether you are a VC or angel investor, in general you need to have an understanding—of cleantech technology, industry dynamics, and the regulatory landscape. So to answer your question, there are lots of similarities, but for cleantech there are unique qualities that are not entirely appreciated. It is still a challenge for people to appreciate and understand it.

How does the CalCEF Angel Fund as a group make investment decisions?

When I put together this fund, I had a lot of experience as an angel investor. I also had a fair amount of experience as a lawyer, employee, and partner with other groups and entrepreneurs. On the venture capital side, I noticed the emphasis on investing in management, not just in select transactions but in the follow-on relationship of supporting and managing and helping the company to grow. I really liked that aspect of venture.

What I liked about the angel side was the collaborative nature and how a lot of really smart people, with a diversity of experiences and knowledge base, come together in a room to discuss aspects of a deal and make investment decisions.

So, when I created this fund, I wanted it to be a little different and I'm not aware of any other venture funds like this. We have limited partners, which are both institutional and individual, and although the ultimate investment decisions are always mine, we analyzed investments through an investment committee that was elected by the limited partners. At each investment committee [IC] meeting, I opened it up to all limited partners. There were five IC members, but we often had up to 12 limited partners in attendance. Everyone participated in what was often a rather lively conversation and debate about a possible investment. Everyone seemed to truly enjoy the process.

Though there were lots of non-voting people there, their voices were equally important. Their background and experience were very valuable, and we always had a couple antagonists or protagonists in there. I loved the conversations that really allowed for deep thinking and critical analysis.

We made an investment in about one out of every 150 companies that applied, so a bit less than the typical venture fund. This was valuable for our limited partners, and it created a fan base for us because they loved our discipline and diligence. They also learned a ton, felt that they were contributing, and that we valued their input. As far as I'm concerned, my LPs are all really smart and I wanted to learn from them. I wanted us as a group to benefit from backgrounds and experiences that I personally don't have.

It was a structure that some of my colleagues who were venture capitalists wrote off: "That's ridiculous. You're never going to be able to manage it. It's going to be a free-for-all," they said. But it never was, not a single meeting. Sure, we got into heated discussions sometimes, but it's about how you manage the situation and how they respect your voice as a general partner. I thought it was one of the most enjoyable, challenging things that I have done in a long time.

Has there ever been an instance where an investment decision changed completely because there were LPs in the room? Either the deal was not funded, or it ended up getting even more funding because the LPs wanted to put in their own money?

I'm sure. I would have never made an investment unless the limited partners were enthusiastic and interested in it. I am very respectful of people who push back; when somebody raised a critical issue, I would do more diligence to find out the answer. If it was not a good answer, we probably didn't invest.

Everybody was very serious about making the right investment decision. Nobody had little off-the-top-of-their-head comments like, "I don't like this." Everybody was very objective and mature about the process, so maybe I got lucky. I always felt that everybody was putting a lot of effort and time into thinking through it and being very thoughtful of their comments.

There was one time where I really wanted to make an investment and the group decided not to and I just could not get it across the line. Then somebody articulated why, with really good points of proof. But I think we all have that experience occasionally. We act based on what we believe is best or we make investments in some things that turn out to be bad. You have to go by what is your best judgment at that time.

The Angel Fund has a 10-year life span. How is that decided, especially given that cleantech companies have a large initial capital cost and might take a longer time to generate a return?

It complies with the standard structure of venture capital investing. We really didn't do much different in the actual legal structure. The only thing that I really changed was in the operational process, using an open investment committee structure. Otherwise, we would look basically like any other venture fund: it's 10 years with two one-year extensions. If the Fund still has active investments at the end of this time period, some funds extend operations by LP vote. Others distribute out the ownership share in the individual companies.

My priority during board meetings is to focus on our portfolio companies, on what is the end gain, what is the exit, and making sure that we achieve that.

With your fund, you are usually one of the first people—if not the first—investing in a company. You mentioned earlier that decision factors are different for early-stage investors. If you had to drill down with just one primary decision factor for investing in a cleantech company, is it the winning attitude, the team, the market, the return, or is there some other particular that dominates over other considerations?

For me, it's a question of order of importance, but all four aspects—the attitude, team, market, and potential return—have to be present or understood. But if I have to select one factor, for me, in energy, it's the market. You have to really understand the market for the product. This is a very challenging industry and market, and you have to fully appreciate not just the current market but also the go-to-market strategy. The questions you need to answer are: Who are the players in the market and what are the opportunities of interacting with them? What are the realities of market pull? Are you really solving a pain point in the market that the market is willing to pay for? And there are so many more questions.

For instance, energy-efficient windows are green. We have shown that we can cut down on energy utilization considerably with a lot of the new technology. In new buildings, they might be incorporated. But for retrofits, trying to replace windows in a multi-tenant building is not going to happen.

It sounds all good, we're saving 40% on energy and all that. It's the same thing with energy-efficient lighting, but the practical matter is that in most of the buildings out there, particularly in the multi-tenant buildings, the owner has no real big incentive to do anything. The only incentive is a possible regulation requiring owners to include green improvements, along with changes to the building that owners want for other reasons, as part of another upgrade. That's why it has taken 30 years to get a lot of buildings upgraded to acceptable standards. Another driving force could be high vacancy rates, which induces landlords to highlight certain attributes to attract tenants, like lower utility bills.

I don't think people appreciated a lot of those factors, and the recession didn't help us. A lot of large companies that incorporated some new technology had been willing to place the cost of new equipment onto their books, but that option was not available during the recession. In addition, companies were looking for a short payback time period of one to two years. These are all complications you face when you're in this industry, and you have to understand that. You have to ask the prospective customer, "Will you actually buy it? Don't tell me that you think it's cool. Would you buy it?"

In the early days, I don't think people knew that you should ask those questions. They certainly did not know the questions to ask.

I am still going to say that market is the primary consideration, but following that, if you don't have good people and good technology, it doesn't matter if there is a market there. The list of priorities might change, but all of these factors are real.

In an interview you did,[123] you mentioned two specific areas that might be worth consideration for investing. Do you still believe in battery and storage technologies? Are there additional areas worth looking at?

I'm still a firm believer in energy efficiency as well. I believe that we can solve a lot of our global warming or energy dysfunction issues by being energy efficient. The state of California has proven the importance of energy efficiency. Since the early '70s gas crisis when the state instituted energy efficiency regulations, the per capita consumption has remained flat for the state of California while that of the rest of the United States has tripled.[124]

When we're talking about other kinds of technology outside of that, what we do not do well yet is battery technology. Battery technology is better than it used to be, but not nearly as good as we need it to be, particularly when we're talking about storage. We are producing renewable energy, but we are not doing a good job at storing it for when we really need it.

There has to be a marriage between renewable energy and cost-effective storage in order to really have the opportunity for rapid renewable energy deployment. Adoption rates are lower than we believe they should be, largely because of storage problems: we can generate the energy but don't have it available when we need it.

We need to find ways of creating the ability to store available energy for the time of highest use. Wind blows more at night than it does during the day. Meanwhile, the highest time for solar is midday rather than the highest level of consumer use between 5:00 and 6:30 in the evening when people get home and crank their air conditioners on or start using their appliances.

It is such a mismatch that way. If we had the ability to inexpensively store energy and use it later, that would allow us to adopt more renewable energy, because it is becoming cost effective. We are starting to commonly talk about the basic success of

[123] Leslie Stone, "Susan Preston, General Partner for the CalCEF Clean Energy Angel Fund," *Opportunist*, 25 July 2012, http://opportunistmagazine.com/susan-preston-general-partner-for-the-calcef-clean-energy-angel-fund/.

[124] Natural Resources Defense Council, "California's Energy Efficiency Success Story: Saving Billions of Dollars and Curbing Tons of Pollution," *NRDC Fact Sheet* (July 2013), https://www.nrdc.org/sites/default/files/ca-success-story-FS.pdf.

alternatives on the grid. The cost is getting lower for solar and even lower for wind. The cost for both is still higher than coal, but we are getting closer. The problem is that we don't have the ability to match up peak renewable production with peak demand: this will be solved with cost-effective storage.

Batteries are an old technology with tons of potential improvements that include reducing the cost of batteries, improving their performance, greater voltage, greater energy density, and power. There are so many different areas in which we could see a lot of improvement, and progress is slow in this industry, so we have lots of opportunity there as well.

When you first picked Alphabet Energy as an investment, what was your investment thesis on the market and the team as a reason for investing in them?

Waste heat recovery[125] is one of those big game-changing scenarios—if you can actually develop a commercially cost-effective technology—because nearly 50% of energy produced is wasted.[126] If we can capture that even to a small percentage, we're saving so much and doing well in energy efficiency.

Alphabet Energy's technology is silicon-based, [127] which is a very cheap material, and therefore it has the opportunity of meeting the fixed goal of producing a sea change in how we look at energy. That is why I love it.

We loved the technology. It is a massive global market, but high risk. When we invested, the technology was only proven on a single nanowire, so it was really early-stage. Matt Scullin, the CEO, struck us as a brilliant scientist, but relatively inexperienced in business. Not only has he grown as a scientist, but he has become an absolutely amazing leader and extremely capable business person. He is one of those rare entrepreneurs who understands both sides of a business, so it's a great investment. It's still a long way from being successful, but it looks really promising in a number of ways.

The company now has a commercial product that is showing great results; a remarkable accomplishment in a relatively short time period.

The other company like that for me is Allopartis, which was one of our initial investments. We actually sold it to a large, global public company. It was sold in 2013, and we received a nice return on our investment. The first time I met the three founders, I knew that it was a company I wanted to invest in because not only was the technology really interesting, but these guys were absolutely fantastic. You

[125] *Waste heat recovery* is the process of collecting waste heat and using it for another purpose, often to generate electricity, in an effort to increase efficiency and decrease fuel consumption costs.

[126] Barry Fischer, "U.S. Wastes 61–86% of Its Energy," *CleanTechnica*, 26 August 2013, http://cleantechnica.com/2013/08/26/us-wastes-61-86-of-its-energy/.

[127] Silicon nanowires have a wide range of applications including photovoltaics, lithium batteries, and sensors.

knew they would work so hard to get across the finish line and were clearly brilliant, approachable, extraordinary individuals, and that sold me. Plus, the market at that time was really big because they were going after the biofuel market.

After that, it changed, but at the time the market was really hot.

What attracted you to Boulder Ionics?

Boulder Ionics is in the ionic liquid electrolyte area. Utilization of ionic liquids as an electrolyte is an emerging area. Ionic liquids give distinct advantages with regard to safety, such as possible fire, because their vapor pressure is negligible.

You read about lithium ion batteries catching fire, which happens because the electrolytes overheat because of their vapor pressure. That's why there are complex cooling systems around them.

Ionic liquids are very different. You can heat them up much higher and they will be completely safe. They also have the potential for higher voltage output. There are a lot of really great things about ionic liquids, but it's a very nascent field. It is an interesting opportunity for incentives and expansion of market, and has really brilliant people and technology.

From my standpoint, this was a bit of a bigger risk for the market, but we saw that the market was increasing for the entrance of ionic liquids and everybody is looking for something that solves a lot of these issues. We are optimistic about the technology, but like any other startup company, it takes time to get to the point where customers are willing to consider utilizing a new material.

How do you work with entrepreneurs in your portfolio? Do you meet on a monthly basis or a quarterly basis? How do you interact with them?

I attend board meetings, and work with many of the entrepreneurs in the interim. Certainly, there are conversations during the board meeting and then calls in between to check in on areas and answer queries to help out. I don't necessarily set a schedule, but because I'm there at all the board meetings it allows for a good touch-point, and interaction.

CalCEF was the founding LP of the Angel Fund, and you have had a number of other LPs ranging from foundations to family offices to wealth managers. Do you think the profile of LPs who are investing in clean energy has changed over time, and if so, why? How has that shifted?

Let me give you an example. We did not seek out another fund. We started talking to all of our individual limited partners, and all of them said, "Yes, please do another fund. We love what you're doing. We love the meetings, and we want to continue investing."

Unfortunately, the institutional investors were hit rather severely by the recession. Their ratios were upside down and venture was not a favored asset class. Among venture, clean energy was definitely close to the bottom. Our institutional investors said, "We're not doing any clean energy. We have been burned so bad, we are not even touching it."

They just had the reaction of, "We are losing money in this so we are getting out." We couldn't really raise funds of a material size without the institutional investors, so we decided not to raise another fund.

I am really regretful of that now because it's a great area, and there are fantastic companies. I think we have pushed ourselves through the heyday of, "Oh cool, new market. Let's go after that one and not really think about what we're doing." Institutional investors are being more thoughtful about it now. The money that's going in is more educated and informed, but it's going to take more time before people start realizing that companies in these areas are being successful and they all start coming back, saying, "Oh, let's get back in."

Do you think there has been a gap between what early angel investors want to fund for cleantech and what VCs want to fund? Is there a gap between those periods of fundraising? If so, how can it be addressed?

I would say there is a gap, sure. Historically, angel investors have been a bit burned by VCs. It's the first money in, it's the highest risk money, and then the VCs come in with unfavorable terms and essentially wipe them out.

It never made any sense to me because I don't see the value in doing it. As a result, I think that a lot of VCs will tend to invest in companies that do not have a huge capital requirement, and in which they can do a couple rounds of financing while having a greater level of control over subsequent terms of investment.

I think that some of the VCs are beginning to realize angels actually do give value. On the whole, angels seem to be getting more sophisticated but we do still have people who get involved as angel investors without proper understanding and training.

Lots of different factors come into play, and it has taken a while for angels to get started, for execs to build relationships with VCs, and for VCs to start realizing that this is a new crop of angels now. This is a new group of people who are educated, have a greater level of experience, are adding value, and can play a critical role in the company's future by being the first, highest-risk group buying in when VCs were not even willing to put their money in.

I think things are changing, but it is still going to take a few more years to hopefully get to a point where we all play nice together.

Has anything surprised you about the energy industry?

The one surprise for me is that many companies would not pay a couple percent more for energy-efficient technology, for things that are good for the environment. I think the energy cost pain is not high enough in the United States; we don't have high-enough energy prices. We don't have federal regulations. We don't have the particular drivers necessary to cause companies to seriously look at efficient and renewable energy as necessary.

I think I was both upset and disappointed that not everyone has even agreed that climate change is a fact, and failing to recognize that fact reduces federal incentives to change energy policy. Overall acknowledgement will be good for the environment, for the economy, and our future.

Of course, that is what is happening in California now, with laws mandating certain steps in the required utilization of renewable energy and the adoption of energy-efficient technology. It's meeting those mandates. I think one of the bigger disappointments is that people don't understand that we have to be stewards as well as make lots of money. I don't like that general misconception, but I understand it.

Insights

Recognize that there will be minimal dollars put toward cleantech investing. Financing cleantech ventures through institutional investors—banks, pension funds, endowments—will be difficult and unlikely for at least a few more years. Institutional investors' wounds are still tender from poor performance across the cleantech market in the early 2000s as well as from the 2008 financial crisis. Some institutional investors have since invested in the market, but they are small in number and are mostly doing so because they have a cleantech background and are more experienced in the cleantech market. Cleantech has fallen low on the investing priority stack. Once companies in cleantech begin to show measurable business promise and generate returns, however, large institutions will be more willing to throw their hats back in the ring.

Inspire angel investors to invigorate the market. Angel investors play a critical role in the next decade of cleantech. As institutional investors are slow to realize that cleantech is entering a new era, angel investors have become even more important to shepherding the market back into a state of vigor. It's a challenging task for angel investors, however: they incur the greatest risks as first investors, and must evaluate a vast majority of unknowns. Their due diligence is more difficult than that of follow-on investors, to identify and prevent risks so that the portfolio company can continue to operate in its early stages, raise capital at higher valuations, and eventually exit for a positive return. Investments in cleantech should of course follow the basic fundamentals of investing in startups—find good teams, strong business models, and understand the market—but understanding the market is particularly important for angel investors. They need to clearly understand whether the founders are solving a problem in the market, and whether people are willing to pay for the solution. Once startups and angel investors can prove themselves successful, institutional investors and traditional VCs will be better able to see the potential value in cleantech investments and make them with less trepidation.

Navigate and adapt to regulation to succeed in cleantech. It's important to navigate stagnant or changing regulation. When policy doesn't change, how does one succeed? When policy moves away from a business, how does it adapt? Many segments of the marketplace are not willing to pay for expensive purchases like energy-efficient products unless they are incentivized or mandated by the government; they are unwilling to buy a product simply because it's better for the environment. In the past, California has proven that regulation can produce dramatic results: with state-instituted efficiency regulations, the state's per capita consumption of energy has been flat, while the rest of the United States has tripled. Regulation will continue to be a hurdle for startups and founders. Startups and investors should keep a close eye on regulations, so that changes do not catch them flat-footed. Business models that rely on a particular regulation should have a backup plan in place in the event that regulation changes.

John Woolard

Former CEO, BrightSource Energy

John Woolard is currently Vice President, Energy at Google where he oversees the company's energy businesses. His first job in the energy business was at PG&E Corporation, after graduating from the Haas School of Business at UC Berkeley in 1997. Seven months after joining the utility, Woolard left and started Silicon Energy—but not before convincing several key PG&E executives to go with him.

Silicon Energy, with Woolard as President and CEO, sold energy-efficiency software to large energy users. After weathering the 2001 financial crisis, Silicon Energy was eventually sold to Itron, Inc. for $71 million.[128] Woolard worked at Itron until 2005, when he left to take a break and think about his next move.

Woolard joined Lawrence Berkeley National Laboratory (LBNL) as a visiting scholar focused on identifying which energy technologies would be successful by 2020, which eventually led Woolard to concentrate exclusively on solar thermal technology. By mid-2005, the potential for utility-scale solar in the United States seemed strong, with high natural gas prices in the wake of hurricanes Katrina and Rita, and state laws mandating that utilities purchase renewable energy. Of the solar technologies available at the time, solar thermal was cheaper on a per megawatt hour (MWh) basis than solar PV. Since utilities generally purchase the cheapest electricity first, plants using solar thermal technology were expected by many to win power supply contracts with utilities.

That same year (2005), Woolard joined VantagePoint Capital Partners as an executive-in-residence. At VantagePoint, Woolard was given the resources to conduct deep due diligence on solar thermal technologies. Based on Woolard's recommendation, VantagePoint chose solar power tower technology for its first solar investment and invested in LUZ, an Israeli company that had commercialized

[128] Dan Gallagher, "Silicon Energy Acquired for $71.2 Million," *San Francisco Business Times*, 21 January 2003, http://www.bizjournals.com/eastbay/stories/2003/01/20/daily14.html?page=all.

the technology by building a series of plants in the United States during the 1980s.[129] The company's name was changed to BrightSource Energy, and Woolard became the company's president and CEO in 2006.

BrightSource's early performance showed that VantagePoint's expectations were well-founded. By mid-2010, it owned 50% of the power supply contracts between California utilities and solar thermal projects—almost 20% of the contracts between California utilities and all renewable projects.[130] The company's projects were expected to produce more than 7.3 gigawatts per year (GW/y), or enough power to supply 1.6 million California homes per year.[131]

Unfortunately for BrightSource, two major headwinds developed in 2010. First, utility-scale photovoltaic (PV) solar panels became price-competitive with solar thermal, mostly due to a massive increase in Chinese PV panel manufacturing capacity. By 2012, solar PV was much cheaper than existing solar thermal technologies, and utilities were not awarding new power supply contracts to solar thermal projects.[132] Some solar thermal developers began converting existing projects under development to solar PV where possible.

The second headwind was specific to BrightSource's first commercial-scale project. Named Ivanpah, the project consisted of three phases located on 3,600 acres of federal land in the Mojave Desert of southern Nevada. Unfortunately, this land is also habitat for the desert tortoise, listed as a threatened species under the Endangered Species Act. In addition to galvanizing a number of environmental groups to oppose the Ivanpah projects, the presence of the desert tortoise caused a more complicated, more expensive, and longer permitting process. BrightSource had 150 biologists working at the site, and its tortoise-saving efforts cost $56 million, including acquiring more land for tortoise relocation.[133]

[129] Solar Reserve (US), Abengoa (Spain), and Sener (Spain), among others, also use this technology. In this book, the terms *solar thermal* and *concentrating solar power* are used interchangeably.

[130] California Public Utilities Commission (CPUC), "RPS_Project_Status_Table_2010_August.xls," http://www.cpuc.ca.gov/ PUC/energy/Renewables, accessed on September 12, 2012. The most recent data published by the CPUC includes data through February, 2016, and can be found at http://www.cpuc.ca.gov/WorkArea/DownloadAsset.aspx?id=9992.

[131] Solar Energy Industries Association, "What's in a Megawatt? Calculating the Number of Homes Powered by Solar Energy," 2017, http://www.seia.org/policy/solar-technology/photovoltaic-solar-electric/whats-megawatt.

[132] Solar PV modules declined to $0.73/W in 2012 from $4.34/W in 2005, or 17% per year. Compare this to the 20-year period from 1985 to 2005, when panel prices declined by 5% per year to $4.34/W in 2005 from $12.17/W in 1985. See Bloomberg New Energy Finance and the Business Council for Sustainable Energy, *Sustainable Energy in America Factbook 2016*, 56, http://about.bnef.com/content/uploads/sites/4/2016/02/BCSE-2016-Sustainable-Energy-in-America-Factbook.pdf.

[133] Debra Schifrin and Donald Kennedy, "BrightSource: Challenges and Prospects for a Concentrated Solar Plant," Stanford Graduate School of Business Case P84, 27 May 2013, https://www.gsb.stanford.edu/faculty-research/case-studies/brightsource-challenges-prospects-concentrated-solar-power-plant.

Given the unprofitable U.S. outlook, BrightSource is now focused on building projects in other markets like China and Israel where government policy or market economics make its technology more cost competitive. Woolard resigned from BrightSource in 2013 after seven years as CEO, remaining as an advisor and board member, before moving to Google to oversee its many energy projects and interests. As a seasoned executive in the energy sector, Woolard shares his unique experiences from both the startup and larger company perspective.

How did you become interested in energy?

I did a Master's in environmental planning first. The environmental planning degree was at UVA [University of Virginia] and the focus was on biodiversity, species issues. That took me into energy and climate change as a big issue. I started to look at climate change as it related to species in biology, which led me into energy. I started looking with a professor at market-based solutions to environmental problems.

After graduation, I did a few things before coming to California in 1995 to get an MBA at Berkeley. I picked Berkeley because it was close to Lawrence Berkeley National Laboratory [LBNL] and I wanted to focus on energy. At the time, there were no energy-focused MBA programs anywhere in the country. In contrast, I lectured at Berkeley recently and about a third of the class was focused on energy—it's unbelievable.

Why did you land at PG&E after business school?

I needed to go live inside of a big entity like that to really understand how the utility side works because it is at the heart of the energy business. My intention the whole time was to start my own company, specifically a software company that focused on energy efficiency. My goal was to be there two years and then leave. I lasted seven months.

What did you do at PG&E? How did you decide to focus on energy efficiency?

PG&E was trying to sell big customers—like Ford, Walmart, and Kaiser—bundled energy services.[134] I was the kid who knew how to run the computer demonstration, so I was in every meeting. Every customer said the same thing: They wanted software but not the bundled energy service. I knew 10-for-10 for a focus group was pretty good.

One guy I worked with had built PG&E's SCADA [supervisory control and data acquisition] system, which is basically the software that allows the utility

[134] Like so many terms in the energy business, *bundled energy services* is a term with many definitions. For example, a utility might "bundle" electricity with additional services such as energy efficiency or energy risk management.

to operate its electric grid. He had done it pre-Netscape, and written a browser because he thought you should be able to use pictures to navigate. He is a great software technologist and he was also the lead architect on the Comcast smart home initiative for PG&E and Microsoft. He had done all that architecture.

He had about 20 years at PG&E, but I convinced him to leave. With another guy, we started Silicon Energy to build software for commercial and industrial customers, to help them with their bill analysis and their meter data management. Basically, we were helping them understand how they could save money on energy.

How did you start building Silicon Energy into a viable company?

First, we went to the VCs. All of them said, "We have no interest in energy. We only do software." So, we came back a few months later and said, "We're a software company that happens to be in energy." We rebranded ourselves and got funded pretty quickly by Nth Power and JMI Equity. We built the company up. From the very beginning, we said we'd never be able to sell to the utilities, so we focused on selling to the commercial and industrial customers. We deployed first to University of Southern California; then we landed Walmart, the U.S. Navy, and Ford Motor Company. We got big customers to buy what is now called "enterprise energy management software," which gives users the ability to measure, monitor, and understand their energy use.

Then the 2001 economic downturn came. Our board was really good; they decided very quickly to size the company to where we did not need to raise a dollar. Within weeks we were cash flow neutral. That gave us options that our competitors didn't have because we didn't need to raise any money.

All of our competitors went under. We just kept building up slowly and methodically, adding customers until we were able to sell the company to Itron for $73 million in 2003.[135] That was the first successful cleantech exit of 2003. It was a good landing for the customers, the employees, everybody. All the investors made money.

When did you leave Itron? How did you become interested in solar thermal technology?

I left in 2005. I took six months off to spend time with my family and then went back to LBNL. I didn't want to go work in venture capital—which was a world I knew fairly well—because it's a reactive world. They don't always think forward very well. I wanted to create my own thesis about where things were headed.

At LBNL, I looked at the macro trends and created this strategic map. I looked at everything from biofuels to wind to solar to efficiency. Wind was fairly mature

[135] Public sources give this figure at $71.2 million. See Gallagher, "Silicon Energy Acquired for $71.2 Million."

from a technology perspective but interesting from a development perspective; you weren't going to start a wind company or find a new wind technology in 2005. That was already happening inside of GE and places like that.

Biofuels didn't add up. They categorically do not make sense—there are too many inputs, from fuel to water. It never added up to me.

Energy efficiency was and is still the smartest thing in the world to do, but I had done that for seven years.

So, I looked at solar. I was talking to the utilities as well because they were the long-term buyers. Utilities are thinking 10 and 20 years ahead. They were telling me that the 20% and 33% renewable targets were real; they were going to have to buy a ton of renewable energy and they were concerned about reliability and intermittency. That led me toward solar thermal.

Why did you reject solar PV at that time?

First, it was really expensive, like $6 per watt. Second, the balance of system [BOS] was expensive. You knew the cost would come down, but how fast and how much? It was too hard to predict. Plus, it is very intermittent, which the utilities won't like. Of course, ultimately the low cost of solar PV won out.

How did BrightSource start?

While at LBNL, I joined the board of the California Clean Energy Fund [CalCEF]. We looked at different venture groups to invest some of the CalCEF money, and ended up with Draper Fisher Jurvetson, VantagePoint, and Nth Power.

Subsequently, VantagePoint asked if I would be an executive-in-residence. I took the position. They gave me the funds to perform deep due diligence on solar thermal.

I looked at every solar thermal group out there. I went back to VantagePoint and said, "There are two companies that are going to work: Abengoa, which is a big company that doesn't need our money and is capable from an engineering perspective, and a company in Israel, LUZ II." There were about six or eight other companies out there, but I preferred to sit out rather than put money into them; they just weren't going to work.

What attracted you to LUZ?

They had built the first nine solar power plants in California. The LUZ team had forgotten more than anybody else knew. They had just seen so much.[136] I flew over

[136] LUZ built nine solar thermal projects totaling 354 MW in the United States between 1984 and 1990. See California Energy Commission, "Solar Energy Projects in California," 2018, http://www.energy.ca.gov/sitingcases/solar/.

there thinking I was an expert, and it was very humbling when you met the real experts. It opens your eyes.

These were legacy people from the SEGS[137] projects?

Yes. They had learned a lot from SEGS projects. They said that they would never build using parabolic trough again, because it was lower efficiency and low temperature.[138] The only way they thought you could win, and I still think this is right, is with a tower-based approach. The optics are better, the losses are lower, and you can get to higher temperature and higher pressure.

Also, you have to ask, what can you get financed? A steam plant can get financed; it has low technology risk compared to molten salt storage[139] or something like that.

How did you move from VantagePoint to LUZ?

The VantagePoint CEO asked me. He was going to pitch the 18 partners inside VantagePoint and he asked to meet with me. I thought it was for coaching about how to deal with the partners, but he said, "No—I'd like to just ask you if you'd want to come run this."

Up to that point, I'd said no to everything that VantagePoint wanted me to run. I wasn't sure if I wanted to go be a CEO again. It's a lot of work. This one made me think, though. I went to the partnership, declared my conflict, and stayed out of the process. It took me four to five weeks to sort it all out, but I ended up deciding to be a CEO.

A side story is that I had helped VantagePoint write the term sheet for the deal. As CEO, I was about to sign the term sheet and thought, "I wrote that damn thing." I had in effect negotiated against myself. So, I went to the partners and asked them

137 LUZ's projects were named SEGS 1 through 9 and are collectively known in the industry as "the SEGS projects"—SEGS is an acronym for Solar Energy Generating Systems. Ibid.

138 The two forms of utility-scale scale solar thermal technology most widely deployed are parabolic trough and power tower. Both technologies use mirrors to concentrate sunlight on a heat transfer fluid that creates steam to feed a steam turbine, which in turn produces electricity. For more, see Sargent & Lundy LLC Consulting Group, *Assessment of Parabolic Trough and Power Tower Solar Technology Costs and Performance Forecasts* (National Renewable Energy Laboratory, October 2003), http://www.nrel.gov/docs/fy04osti/34440.pdf.

139 An evolution of the solar power tower design is the addition of storage. Molten salt is the fluid in the tower that is heated by sunlight; rather than use this heated fluid to immediately make electricity, it can be stored (with some heat loss) and used to make electricity later. This setup gives the power plant more operating flexibility; for example, it can generate less electricity when demand is low (and the sun is shining) and make more electricity when demand is higher. SolarReserve was the first deploy this storage technology in its 110 MW Crescent Dunes plant in Nevada. See Knvul Sheikh, "New Concentrating Solar Tower Is Worth Its Salt with 24/7 Power," *Scientific American*, 14 July 2016, https://www.scientificamerican.com/article/new-concentrating-solar-tower-is-worth-its-salt-with-24-7-power/.

to remove the terms that were egregious. The partners were great people; it was a 10-minute discussion.

Were there other investors?

I called one guy, Raj Atluru, at Draper Fisher and they came in. He's just a good thinker about the energy space. That was it. We brought in around $16 million in September 2006. We changed the name to BrightSource. That was the beginning.

How did you put together your team?

I built a network of people I met along the way. If you have a world-class network, you may be able to hire 80% of the positions from your network. This method is more successful than going through recruiters and getting randomized input.

What was your three- to five-year vision for BrightSource?

The vision back then was actually more modest than the total number of PPAs [power purchase agreements] that we ended up signing. We thought we would do 100 MW, then 200, and eventually build up to a few hundred MWs per year. Then we started talking to the utilities, and they had a big appetite. We ended up getting overwhelmed. We were at a price point that was lower than anybody else, and ended up signing 2.6 gigawatts of PPAs from the utility RFPs [requests for proposal] between 2007 and 2009.

How did you determine what price to bid to the utilities?

Our cost was 70% cheaper than parabolic trough and 50% cheaper than solar PV at that time. And knowing that we were cheaper, the question became how high could we bid and still win? You don't want to leave money on the table.

So, we bid a price that was both just below what we thought would be the lowest price for solar trough and below the market price referent [MPR].[140] We thought that if you bid above the MPR, the California Public Utilities Commission [CPUC] would never approve the deal and then you'd never get the project financed.

For example, let's say the installed cost, operation and maintenance costs, and financing package resulted in needing a contract price of $100/MWh. A PPA at that

[140] To evaluate the cost of power supply contracts submitted by investor-owned utilities for approval, the California Public Utility Commission (CPUC) is required by law to establish an estimate for the long-term price of electricity. This long-term price is called the *market price referent*, or MPR. The calculation of the MPR is complicated and controversial. In theory—and, mostly, in practice—a power supply contracted priced below the MPR is more likely to be approved by the CPUC than a contract priced above the MPR. See California Public Utilities Commission, "Market Price Referent (MPR)," 2018, http://www.cpuc.ca.gov/General.aspx?id=5599.

price level is financeable. The price of solar trough and the MPR were at $130 or $140. So, we would price at $125.

The result was that we were able to get PPAs for a couple of gigawatts at those prices.

With several gigawatts of power purchase agreements in hand, what was the plan to build out the underlying projects?

We knew you had to get the first commercial-scale project up and running before building the next one. In our minds, the first plant needed to be running for one year before we could get financing for the next plant. So, we planned on staggered construction and financing, and built this timeline into our PPAs.

When did you start building the pilot or demonstration project?

We started immediately. Demonstration is a step that was missed by a lot of people who hadn't been around energy-project financing[141] before. One thing we did well is we always put ourselves in the lender's shoes: What would I need to do to lend money for this?

The most important constituent is the independent engineer for the debt providers. If the independent engineer doesn't believe in the technology, then you've wasted all your time and effort, because there is no way you will get funding for a big project. Some companies try to skip this step.

So, the independent engineer is the customer. We would drive that point home to our people every day. The independent engineers helped us develop the specifications for the demonstration facility, and then we built it on our own balance sheet. It took us 18 months to build the 6 MW pilot plant start to finish, and it performed at 104% of design capacity.

Was this process of building the pilot project running in parallel with other efforts?

Yes, we were also negotiating with the PPAs with the utilities, we were negotiating with the EPC contractors,[142] and we were finding tax equity investors. All of these things were building toward project financing.

[141] *Project financing* is a method of funding a large, long-term infrastructure project in which the debt raised is repaid from the cash flow generated by the project after it is completed. Project-financed debt is non-recourse; in the event of default, lenders do not have rights to the borrowing company's other assets.

[142] *EPC* stands for engineering, procurement, and construction. Basically, developers hire an outside firm with the expertise and resources to build the plant. The contractor—Bechtel in this case—becomes responsible for building the plant by a certain date and for a certain price, and guarantees that the plant will perform as designed (or else pay damages). Without a reputable, financially sound contractor like Bechtel as EPC contractor, BrightSource would have been unlikely to find investors.

How did you find your ultimate EPC contractor?

We had bids from multiple EPC firms and had short-listed two. From those bids, we knew what the cost would be within a range. We hadn't done a final selection because once you do that final selection, you give up all of your leverage. The relationship goes from "I've got you" to "you've got me." So, we waited. We negotiated every single term of the EPC contract in excruciating detail with both parties, so that neither would have the ability to hold us hostage over the negotiation of any term of the contract.

Ultimately, we picked Bechtel. Then we hired an ex-Bechtel lawyer to be general counsel. If you're going to have Bechtel as your EPC counterparty, you better have people on your team that have worked there and know how they think.

How did you find tax equity investors?

We had tax equity from an investment bank; this was pre-Lehman Brothers when the whole world was long tax equity. So, we had that problem solved.

Then Lehman Brothers happened, the world fell apart, and everything was crashing.[143] The bank investing for tax equity lost all of its taxable income for 5 to 10 years, and I think the bank even closed its tax equity group. The project debt market also shifted dramatically.

Were you looking for debt when Lehman Brothers collapsed?

We had just started discussions with debt providers when the financial collapse hit. Then there was no way to get project debt post-Lehman except for the DOE loan program.

We were fortunate because we had been looking at the loan guarantee program all along the way. Some of our folks thought it was a waste of time, but I thought it was a low-cost back-up plan for us. A guy on our team took charge of the loan guarantee process, and he stayed focused and ignored the people who said he should work on something else. So, we made sure we could participate in that program and we made it through every gate to the "Sweet 16," which was the first group to receive the loan guarantee.

What is your opinion of the DOE loan guarantee program?

Its roots are in nuclear plant construction, and then the DOE added renewable

[143] Broadly, Woolard is referring to the financial crisis that began in 2007 by the collapse of the housing bubble. Lehman Brothers—a storied investment bank—filed for bankruptcy, J.P. Morgan bought Bear Stearns, and Bank of America bought Merrill Lynch. During this time, as the various banks struggled to survive, they effectively stopped lending to or investing in renewable projects (or, really, anything else).

project financing and manufacturing. Obviously adding manufacturing was a bad idea; experienced energy people at the time knew that guaranteeing loans on manufacturing plants was a bad decision. Venture risk—picking technologies—is much different than project risk. It was a crucial mistake driven by the desire of policymakers to create jobs.

However, the loan guarantee underwriting process was very thorough and harder than the process at commercial banks. The DOE retained very good outside engineering expertise. As a taxpayer, I felt good about the diligence process. It was really serious. There were delays, but any financing is going to get delayed a little bit.

Were their other causes of delays?

There were permitting delays caused by environmental issues, mainly because I think Ivanpah was the first project built on federal land. The main environmental issue was the desert tortoise.

Ironically, the site was not good desert tortoise habitat. It had one-fourth the tortoise population density of healthy habitats. The site was near a highway, surrounded by transmission lines and gas pipelines. Site access was through a chain link fence behind a casino.

But Ivanpah was the first project to be built on federal land and that made it high-profile. Also, labor and other groups can use an issue like the desert tortoise for leverage to meet their goals; once they get their project labor agreement, their concerns about the environmental issue go away. People know about this practice, but nobody likes to talk about it.

Despite these delays, did BrightSource have sufficient working capital? Was Ivanpah hitting its project milestones?

We had enough working capital. We kept our overhead low, and we continued to focus on hitting the milestones to get project financing for Ivanpah. We had raised Series B and C before Lehman. Series D was post-Lehman, and it was a tougher raise; we got money from outside the United States.[144]

[144] After many delays, BrightSource's Ivanpah project reached commercial operation in January 2014. The company's other U.S. projects also experienced delays, and ultimately it lost all but two of its remaining PPAs with California utilities. Prospects for additional U.S. projects are grim. Power-tower technology is no longer price-competitive with other generation technologies, including solar PV, and most utilities have already met their current and future renewable targets mandated by state policy. Nonetheless, there is a bright spot for Brightsource. Many international markets have higher electricity prices than the United States, and the company may be able to compete in some of these markets. Brightsource has announced the financing for a new project in Israel, and the establishment of a joint venture to pursue projects in China. See www.brightsourceenergy.com/projects.

What are some key lessons you have learned?

To have debt capacity to build any large plant, any project that costs more than $1 billion, you need something like the U.S. government to provide a loan guarantee. Take the Shepherds Flat wind project.[145] Without the loan guarantee, that project isn't 800 MW, it's 200 MW. If it's just the private sector doing the financing, you're going to have smaller projects.

If you do the math, to stop at 450ppm[146] you need 80% carbon-free energy sources by 2040 or 2050; to do that, you need to build 1 GW *per day* of zero-carbon emission power plants. Take all the solar PV built in the United States and we have maybe 30 GW, so we're not even coming close. I don't think anyone has digested the disconnect between the problem of climate change and the pace of adding carbon-free power plant capacity. So, if we're trying to solve the climate change problem, we need big tools that can be used on a big scale.

Another lesson: don't manage your energy policy through the tax code, because it sends false price signals. This means you build generation without acknowledging how expensive it really is. Despite what some people say, solar PV isn't cheap. When you remove the subsidies, solar PV costs more than ten cents per kilowatt hour, which is expensive.[147] Tax subsidies hide this fact, which means that excess private capital is invested inefficiently and the taxpayer is paying more than necessary.

On top of that, you have the stop/start nature of tax subsidies. The worst example is the Production Tax Credit for wind; every time it nears expiration, new capacity additions drop by 30% to 50% the next year. A simpler way is to use renewable portfolio standards.[148] Utilities have no choice but to purchase renewables regardless of whether there are tax credits. The utilities just buy the energy at a higher price.

[145] Located in eastern Oregon, the 845 MW Shepard's Flat project is one of the largest wind farms in the United States or the world. The Department of Energy provided a partial loan guarantee for the $1.3 billion the project borrowed to get built. See U.S. Department of Energy Loan Programs Office, n.d., "Shepherds Flat," http://energy.gov/lpo/shepherds-flat.

[146] In theory, if the concentration of CO_2 is limited to 450 parts per million (PPM) then the increase in global average temperature would be limited to 2°C. Above 2°C, the probability of catastrophic climate-related events goes up. There is debate—beyond the climate-science deniers—about all of this. For example, some scientists think only a 1°C average temperature rise is allowable. As of this writing, CO_2 concentration is above 410ppm. NOAA Earth System Research Laboratory, "Up to Date Weekly Average CO_2 at Mauna Loa," accessed 6 April 2018, http://www.esrl.noaa.gov/gmd/ccgg/trends/weekly.html.

[147] Bloomberg Energy Finance estimates that 2015 solar PV prices, ex-subsidies, range from approximately $.07/kwh to $.20/kwh with an average of $.12/kwh vs. on-shore wind average price of $.08/kwh, and natural-gas combined cycle of $.07/kwh. See Bloomberg Energy Finance, *Sustainable Energy in America Factbook 2016*, 35.

[148] *Renewable portfolio standards* are a regulatory mandate to increase the production of energy from renewable sources. It's also known as a renewable electricity standard.

People have difficulty understanding that someone pays the subsidy—it's either the taxpayer or ratepayer. In Europe, the ratepayer pays the subsidy through a high electricity price. In the United States, the taxpayer pays most—at least 60%—of the subsidy and the ratepayer pays the rest.

In the case of distributed solar PV,[149] since it uses the grid as a back-up system, both the taxpayer and ratepayer pay. Rooftop solar PV is about 80% subsidized: 30% from the Investment Tax Credit, 20% from the accelerated depreciation, and then 20% to 30% is cost of the electricity grid back-up.

So, by creating artificially low prices for renewable energy through the tax code, you send false price signals, and uneconomic generation is built. It's just not efficient.

Beyond tax code, what is a takeaway for you?

I don't think the venture community understood the energy space. At BrightSource, I would have meetings with very reputable Silicon Valley investors, and by the end of the meeting they hadn't asked the questions that were the most fundamental to the business. This means they didn't understand the risks. As an entrepreneur, you don't want investors who don't understand your business and what the risks are.

Were these investors aware that they were uninformed?

No, they thought that they were informed. I think people who have been successful investors elsewhere develop a level of self-confidence in their investment abilities.

You also had a period when too much money entered the cleantech sector. Billions flowed in and there weren't enough good deals, so money went into deals that weren't very good. As the investor, you either invest more in each company, which increases valuation and expectations, or you invest in deals that you would not have otherwise picked. That's how you turn tens of billions into a few billion.

Insights

Understand market pull versus technology push. Before starting Silicon Energy or convincing VantagePoint to buy BrightSource, Woolard knew what the market wanted. With Silicon Energy, he knew the customer wanted software, not bundled

[149] *Distributed solar*, or *distributed generation*, is the use of smaller-scale power generation technologies located close to the building or facility being served. In the case of rooftop solar, the homeowner uses electricity from the power grid at night or when it's cloudy, when the PV cells are not generating electricity.

energy services. With BrightSource, he knew from talking to utilities (mostly in California) that they would have significant future demand for renewable power supply and that solar—compared to wind or other renewable technologies—would be part of this supply. Knowing what the market wants before building or selecting the technology was key to the success of Silicon Energy and the early success of BrightSource.

In commodity markets, price matters. A common saying in the electricity market is that "an electron is an electron." The primary goal of an electric utility is to provide reliable electric service at the lowest cost. Practically, this means that utilities use the cheapest electron first. Knowing that utilities would have to buy solar was only part of Woolard's business case. He had to find a reliable solar technology that offered the lowest cost per MWh. In 2004–2005, solar power tower technology was the cheapest. The 2.6 GW of supply contracts won by BrightSource prove this point.

Focus on project financing. Because BrightSource planned to build and operate multi-billion-dollar power plants, they knew that they would need project financing. They developed an understanding of who likely investors would be, and what these investors needed in order to make a commitment. For example, debt investors require an independent engineer's assessment of the technology. To ensure the likelihood of a positive assessment, BrightSource began working with independent engineering firms early, and made sure that these firms had input in the specifications of the demonstration plant.

Take advantage of government programs. Over objections from other team members, Woolard had the company apply for the Department of Energy's loan guarantee program for its Ivanpah project. While the company expected to finance this project through traditional debt and equity markets, the financial crisis that started in 2007 effectively closed—or at least severely limited—the capital markets for projects like Ivanpah. Had the project not received the loan guarantee for $1.3 billion of debt, it is unlikely that Ivanpah would be built and operating now.

Find investors who know the energy business. In Woolard's experience, venture capital investors did not understand the energy industry and therefore did not understand the risks.[150] Moreover, his view is that many of these investors—

[150] In 2013, we took i3/Cleantech Group's (www.cleantech.com) list of the top 20 cleantech investment firms since 2003 based on total dollars invested and looked at the biographies of each person listed on each firm's website for energy industry experience. We defined "energy industry experience" as anyone who had either worked for an energy company or provided services to an energy company (like management consulting or investment banking). We found that fewer than 20% of the people working at the leading cleantech investment firms had any energy industry experience. While this evidence is anecdotal—firms experience turnover, website biographies may

who were successful in tech and other industries—did not realize that they did not understand the energy business. This lesson aligns with other interviews in this publication: lack of understanding about the energy business, combined with too much capital to invest in the energy sector, created a recipe for the dismal investment returns by many cleantech VCs.

Tax subsidies are an inefficient policy tool. By using the tax code as a tool of energy policy, power generation can be built that would otherwise be uneconomic.[151] This means that private capital is invested inefficiently. Subsidies, regardless of their form, necessarily distort price signals—however, policies made can reflect the political realities, often messy, at that time. In case the case of tax subsidies, some argue that the tax code was the primary policy tool available to clean energy proponents. Still, a less efficient tool may be better than no tool at all.

be incomplete, firms use outside consultants with industry expertise—it did make us wonder how these firms were making cleantech investment decisions. And, ultimately, was one factor in the decision to write this book.

[151] Some readers may be thinking, "But what about oil and gas subsidies?" Fossil fuels and nuclear plants receive tax subsidies as well, which also distort price signals. A 2015 CBO report estimates the tax cost of energy subsidies are $2.7B for energy efficiency, $7.7B for renewable energy, $4.7B for fossil fuel, and $0.2B for nuclear power. Congressional Budget Office, *Federal Support for the Development, Production, and Use of Fuels and Energy Technologies*, November 2015, Figure 2, https://www.cbo.gov/sites/default/files/114th-congress-2015-2016/reports/50980-energysupportonecol-3.pdf.

Tom Baruch

Founder, CMEA Capital

Tom Baruch is well-known as the founder of Chemical and Materials Enterprise Associates (CMEA) Capital, an affiliate fund of New Enterprises Associates (NEA) now known as Presidio Partners.[152] CMEA was founded as a $40 million fund with corporations and financial institutions as limited partners in 1989. The goal was to invest in startups focused on specialty chemicals and materials technologies. CMEA became independent from NEA in 1997, and as of 2015, the firm had raised eight funds and held $1.2 billion under management. As part of CMEA, Baruch has invested in Aclara Biosciences (acquired by Monogram Biosciences), Netro (NTRO), Entropic Communications (ENTR), Flextronics (FLEX), Symyx Technologies (SMMX), Silicon Spice (acquired by Broadcom), Codexis (CDXS), and Intermolecular (IMI), among others. Baruch has been a partner emeritus at CMEA since 2010.

Baruch's experience in venture investing began at Exxon Corporation in the 1970s. At the time, the firm was seeking to diversify from oil and gas and wanted to focus on new types of energy devices. In his 12 years at Exxon, his investments included companies creating some of the first lithium ion batteries, photovoltaics, and those in the electronic materials sector. He began his career in 1965 as an engineer at Battelle Memorial Institute, an R&D company based in Ohio, and later started Superform Metals, a California-based company that supplies products to the aerospace industry.

After CMEA Capital, Baruch became an investment partner at venture capital fund Formation 8. The firm raised a first fund of $448 million in 2011, and then

[152] In April 2015, CMEA Capital unveiled a new corporate identity as Presidio Partners, with a diversified investment strategy focused on information technology, life sciences, and energy technology. See "Presidio Partners Announces New Fund, Unveils New Corporate Identity, and Announces Additional Senior Management" (news release), 21 April 2015, http://www.prnewswire.com/news-releases/presidio-partners-announces-new-fund-unveils-new-corporate-identity-and-announces-additional-senior-management-300069110.html.

decided to not raise a third fund in 2015. Limited partners included the founders of PayPal, Palantir, Yahoo, Yammer, conglomerates in Asia, and institutions in New York.[153]

Baruch is also the founder of his family fund, Baruch Future Ventures (BFV). The fund's capital consists of his personal wealth, and the average investment ranges from $25,000 to $1 million. Recognizing that Earth's population is on track to exceed its available resources, BFV invests in teams working to make life on this planet sustainable.

Over his career, Baruch has also served on a national advisory committee on innovation and entrepreneurship, and advised ARPA-E under the Obama administration. His experience as an investor for over 40 years puts him in a particularly good position to comment on the rise and fall of cleantech.

How did you become interested in venture capital in the first place?

I was working at a company called Battelle Memorial Institute as a junior engineer relatively fresh out of engineering school, and I started doing more work related to patents and innovation. Early in that period, I became exposed to work Battelle was doing on a novel device: a copying machine that had been invented by Chester Carlson in Rochester, New York. His company was called Haloid Corporation [now Xerox].

Carlson had a lot of problems with that device, which needed the application of some advanced materials technology. Battelle had that technology because of all the work the company did for the government and the Atomic Energy Commission. I had the opportunity to work on some of the patents related to the difficulties with the copying machine, which is plated with selenium.

At one point, Haloid could not pay Battelle for their services, so they asked Battelle to take stock. There was a lot of discussion, and a lot of people were not enthusiastic about that idea because they felt this device would never have a market. It is kind of analogous to the IBM story[154] where Thomas Watson said, "I think there is a world market for about five computers." Many people at Battelle believed

[153] Josh Constine, "Formation 8 Raises its First Fund of $448M to Plug Silicon Valley Startups Into Asian Conglomerates," *TechCrunch*, 18 April 2013, para. 5, http://techcrunch.com/2013/04/18/formation-8/. In 2015, Formation 8 split into two funds, Formation Group and 8VC; see Dan Primack, "What Caused Formation 8 To Split, and What Comes Next?" *Fortune*, 10 November 2015, http://fortune.com/2015/11/10/what-caused-formation-8-to-split-and-what-comes-next/.

[154] This statement by Thomas J. Watson Jr. of IBM is often misunderstood and misstated. When taken in context, Watson was speaking in 1953 at a shareholders' meeting about a new product IBM had been planning. Watson was reporting on market feedback for the planned product. Instead of the 5 orders out of 20 meetings he was expecting, instead the firm received 18 orders out of 20 meetings. Fred Shapiro, "Our Daily Bleg: Did I.B.M. Really See a World Market 'For About Five Computers'?" *Freakonomics* (blog), 17 April 2008, http://freakonomics.com/2008/04/17/our-daily-bleg-did-ibm-really-see-a-world-market-for-about-five-computers/.

that the world did not need a copying machine. They thought, "What's wrong with mimeographing? What's wrong with carbon paper?" I knew there was something with carbon paper-like technology.

When Xerox went public in 1961—it was a big success. Battelle made a huge amount of money in 1960s terms. It solidified for me what innovation could do to create value, especially in technology and the application of technology to a device that, as it turned out, had a very large market opportunity.

In retrospect, people clearly had pain around making copies—they just did not want to admit it. It was a risky but tremendously courageous move for whoever it was at Battelle that advocated that program. It was a great program.

I left Battelle and started a small company in California supplying products to the aerospace industry in 1968. After a couple of years, we sold the company, and then I was recruited to Exxon to get involved in its diversification activity—with a heavy emphasis on materials science, which was always my interest.

Exxon became your first experience in venture investing—would you say your interest and involvement in energy investing primarily stems from materials science?

Yes, I would say so. Exxon's motivation was to create a new business that Exxon could invest in because, oddly enough, at that time they did not think there was much future in the oil and gas business. I was looking for opportunities where you could make a big impact. What I ended up doing was mostly working in areas of electronic materials. I was able to track the evolution of the semiconductor industry from the invention of the microprocessor, which was in the late 1960s, early 1970s.

For example, we funded a company called Zilog, a fantastically innovative company that was going to take on Intel in the microprocessor space. Actually, one of the principals of that company was Federico Faggin, who invented the microprocessor at Intel and then went on to Zilog. His first product was the Z8 microprocessor, which is still used today. It is an amazing device.

I learned a lot about energy investing at Exxon because we did a lot of work in new types of energy devices. We were on the ground floor with things that are big today: we worked on some of the first lithium ion batteries and some of the first zinc air batteries. Batteries are very materials-intensive. The problem with most battery systems is that the discovery process requires black magic and art. It is a conjunction of materials science and know-how.

We also did solar and solar thermal photovoltaics at Exxon; everything obvious, none of which was economic. Then, of course, oil dropped to $10 a barrel. Exxon decided that it should stick with the oil and gas business and disassemble most of the things we did.

Some of the companies continued on successfully, though. One of the companies I invested in at Exxon, Supertex, was a power semiconductor company that went public around the time I left Exxon; it is still doing reasonably well. There is some staying power to these electronic materials companies.

At the same time, I have seen other companies that have become obsolete, like Sun Microsystems. Silicon Graphics was a big hit for NEA, and is now a shadow of its 1980s high-flying self because they were not able to retain that entrepreneurial fervor. They became big companies managed by inexperienced people and outlived their usefulness. Their boards of directors did not have the courage to make the changes that had to be made, so the companies died.

What other sectors have caught your investing eye in the past?

I have invested in all sectors. During the 1980s, I primarily invested in what you might call electronics and semiconductors; I founded CMEA Capital in 1989 and focused on electronics and semiconductors. We had a unique fund—about 50% of our LPs were corporations and 50% were financial institutions.

We wanted CMEA to be known as a strategic fund that balanced strategic inputs with financial gains. It worked out really well. We ended up with $41 million in the first fund, which seemed like a lot of money at the time. We invested over a period of six or seven years in collaboration with NEA. In fact, we were actually considered an affiliate fund of NEA.

The emphasis was on electronics, and some biomedical and biotechnology. We were looking for companies differentiated through leveraging materials science. We were also looking for products and customers who could benefit from a value proposition based on materials science, but we learned quickly that materials science deals are cures in search of a disease. Although we did invest in a few materials science companies, we were always searching for a customer who had some pain and an application that made sense in light of that pain.

Materials science was moving forward dramatically, primarily in terms of electronic devices and especially semiconductors. We also looked into areas such as composite materials and carbon fiber composites.

Although we were investing in a fertile area and had unique domain knowledge at that time, we were very careful, because neither my partner nor I had any substantial financial venture capital experience. My time with Exxon had been mostly strategic, and my partner had been with a corporate venture group as well. We invested carefully and slowly, and looked to create a portfolio with well-distributed risk.

In 1997, we raised our second fund with a similar focus as the first fund, again with corporate limited partners and financial institutions. One of our corporate limited partners in the second fund was Dow Chemical Company. At some point in '98 or '99, Dow, who was interested in working with a firm that had a dedicated life sciences fund, approached us.

Our third fund was initiated by Dow. It was a $40 million life sciences fund with Dow as the sole LP. We mainly invested in life-sciences tools companies. During that period, the first human genome had been mapped; there was a lot of interest in the information that would arise from it as well as the possible breakthroughs in drug discovery and disease treatment. We liked that idea, but because of our affiliation with NEA, which was a substantial investor in biotech, it was clear that such things took too long.

It did not make sense for a small fund to take on all the risk by investing in drug discovery early on and then getting blown out if the drug failed a Phase III clinical trial 10 to 15 years later. By that time, it would be more than likely we would have washed out anyway because we could not keep up that time-delayed pace of investing. We had one or two investments that were like that and they didn't work out too well.

The tools investments we made were really good, because it was a time when the research was expanding significantly, and we could apply a lot of materials know-how and materials science to making tools that were more productive. We ended up investing in a number of tools companies. One of them was a company called Aclara Biosciences, which went public and was a very good return for us. It was one of the first companies involved in the use of microfluidics for doing various kinds of biological research. It was a very innovative company.

That is the story from the end of the 1980s until 2000. That was the life science activity. It was a good time to be investing in both life sciences and electronics.

Focusing on 1989 and when you founded CMEA Capital with NEA, can you describe how that came about?

Dick Kramlich, Chuck Newhall, and Frank Bonsal founded NEA in 1977. It was the first venture capital firm that was bicoastal; it had an office in Baltimore and an office in San Francisco. Dick ran the West Coast office and Frank and Chuck were in the East Coast office. One of their limited partners was 3M, which became an investor in almost all of the first NEA funds. The vice president of research at 3M approached NEA with the idea that NEA might want to focus on applications of materials technology.

NEA had a meaningful payout from a company that had done some work in materials technology, and they were enthusiastic about the idea. One thing led to another, and 3M was able to find other companies willing to collaborate in that concept. Ironically, one of them turned out to be Exxon Chemical, which was a subsidiary of my former employer and also a part of Exxon that I actually did a lot of work with, so that was comfortable.

I would say we were the first venture fund that had a mixed corporate and institutional limited partner base. The concept was to enable our limited partners to get visibility into technology applications outside of their own companies. At the same time, we could get access to their libraries of knowledge in terms of due diligence and possibly for collaborations over the long term. That was a very clever idea.

In some cases, it worked, and in other cases it did not work so well. I would say, on average, it worked. We had some reasonably satisfied limited partners. 3M, for example, came back and invested in several of our funds. Dow Chemical invested in all of our funds. We had a great relationship with Dow, and they ended up investing directly in a lot of companies.

Ironically, we did not invest in most of the companies that our LPs invested in. Our LPs wanted to invest in companies that had a product line, but not a monolithic platform that could grow. These kinds of companies were strategically beneficial for an LP like 3M's core business, but not something that could independently grow into $50- or $100-million businesses.

So, there were some ancillary benefits to the mixed corporate and institutional LP venture fund model, but the corporate limited partners just did not have the staying power. Whoever had the idea originally was transferred to another job and somebody else was put in. So, the model was abandoned. I like that model though.

At CMEA Capital, how were investment decisions made? Were decisions more about group consensus? Were they made at the individual partner level?

The original fund, CMEA Fund I, was a collaboration with NEA. It had three original general partners: NEA, Don Murfin, and myself. Each entity had a vote on investments, and the vote was a majority.

What was really neat for someone like myself, who had just entered this business, was that our partner meetings were held in conjunction with NEA's partner meetings. We would get to hear all the deals that NEA was looking at, all the discussions, and even the reports on companies. If a company had a board meeting, they would talk about what happened there.

It was an incredible learning experience. I could not have had a better learning experience at any other place. I am incredibly grateful for the education that NEA gave me. I just cannot think of anything that would match it.

In CMEA Funds II through VII, NEA was no longer a general partner, and we had a more traditional approach to the world. My recollection is that at some periods we had unanimous decisions among the general partners, and at other times we had a majority.

Both decision styles have advantages and disadvantages. What I do not like about unanimity is that it tends to become too political. There is a lot of the "You scratch mine, I will scratch yours" kind of thing. I think a majority vote is probably a better way to work things out in terms of making investments.

Generally, my experience has been that if you have to talk about an investment for too long, it is probably not a good idea. I know of situations in which we took six months to make up our minds, and then the company was a disaster. It did not improve the outcome to spend more time on discussion; it actually only hardened people's opinions.

You have shared that CMEA Capital was primarily focused on biotech and information technology. CMEA Funds IV and V were definitely biotech and IT. CMEA Funds VI and VII were those two plus energy. What was the investment thesis around energy for Funds VI and VII?

The investment thesis was very similar to the investment thesis we had when we launched CMEA. Materials science is probably the biggest lever you have to pull to improve the efficiency of an energy device, whether it is for storing electrons or creating electrons. That is what energy is about—creating electrons, storing electrons, and moving them around. The materials science environment in which that is done is a critical path area—and it can be leveraged because energy can be produced, stored, and transported cheaper.

You can ultimately make a substitution in the marketplace, but it is not easy because many of the existing energy technologies are highly scaled. There is a certain efficiency, there is a certain amount of "better, faster, cheaper" that just comes with scale. That is why the oil and gas business will never go away: it is so scaled that it is hard to compete if you are starting from one device or platform. You have to have something very robust if you are going to take on a business at that scale.

I would say wind has been reasonably successful. Solar has definitely been successful—not necessarily for investors, but for humanity. Humankind is definitely benefitting. The solar business is booming right now, but nobody makes much money. Some of the people who make money are those who provide services, financing, and generally, clever business models. Primary manufacturing is a disaster.

One of your investments in solar was Solyndra. How did you come by that investment?

Solyndra was in the category of very innovative materials science. I was intrigued by it, because nobody else had come close to this. We might say their geometrical pattern enabled the device to capture more sunlight and obtain a higher level of efficiency and productivity—better, faster, cheaper.

The fundamental materials technology for solar is called copper indium gallium selenide [CIGS]. People at the National Renewable Energy Laboratory [NREL] in Colorado had worked with this material. The founder of Solyndra, Christian Gronet, used that as a launching point for his own work in CIGS. He arranged the CIGS in a geometric configuration of multiple tubular solar-capture elements that enabled them to get more output for the input, which is the name of the game in solar. Most people express it as "efficiency."

Because of this radical new architecture, the Solyndra device could be placed into the field at a much lower cost than devices with conventional silicon photovoltaics. The cost of getting the devices into the field is called balance of system [BOS] cost.

Part of the system cost is in the solar capture device itself, and another part of the system cost is its installation on a rooftop or wherever. The Solyndra device's architecture meant we could actually demonstrate that our BOS was a lot cheaper than those of other technologies. It was a very clever idea. The founder was a very bright guy who came from Applied Materials.

The prices were such that at certain volume—which was attainable—we could make money. For a while we could not produce enough devices. Then as the production problem started to get ironed out, we could not make them at a low enough cost to be price competitive with the Chinese product.

One of China's five-year plans had staked out solar as a primary growth area, which basically meant selling below cost and plundering markets,[155] and that is what they did. During that period, Solyndra was trying to scale production; it just could not keep up with the price declines that the Chinese were putting on their products.

Solyndra was very close to success—but a 10% price reduction per quarter is difficult. It was a tough time. In 2009 or 2010, the price of solar in China was just too low and dropping every day. The market was just pulled out from under us; there was no way we could sell our product at that price.[156] The price kept dropping below our direct cost, which meant we would be giving money away, and you cannot do that. It was getting away too fast.

In retrospect, we should have done things differently, but you also have to consider the atmosphere of the times. We were able to raise money for the company at huge valuations in excess of a billion dollars. Investment bankers were dangling numbers in front of us. When we made the decision to take the company public, every investment banker in the world wanted in and quoted prices that were just unobtainable.

Of course, we did not know it at the time. The numbers sounded logical. Analysts went through the spreadsheets and did the analysis. They said, "You want to sell so much this year, so much next year. You're going to cost this much this year, this much next year"—the whole story. It all sang beautifully, like a church choir, but for many reasons, it just didn't roll out that way.

[155] Wiley Rein LLP, *Summary of China's 12th Five-Year Plans Relating to the Solar Industry*, Coalition for American Solar Manufacturing, 2 May 2012, http://www.americansolarmanufacturing.org/news-releases/chinas-five-year-plan-for-solar-analysis.pdf.

[156] For a discussion of the drop in solar prices in China between 2009 and 2011, see Yingzhi Yang, "Why Millions of Chinese-Made Solar Panels Sat Unused in Southern California Warehouses for Years," *Pacific Standard*, 30 June 2015, http://www.psmag.com/nature-and-technology/why-millions-of-chinese-made-solar-panels-sat-unused-in-southern-california-warehouses-for-years; Kevin Gao's presentation at the 2011 International Solar Energy Technology Conference offers a detailed view of the prices and costs of solar in China and the United States in 2010: "U.S.A. and China Solar PV Market & PPA Solar Project Development," 27 October 2011, http://www.isetc.org/English/Archives/201110/ISETC2011_Kevin_Gao.pdf.

I think the world has fixated on the loan aspects of what happened at Solyndra,[157] and that's what I cannot talk about. Frankly, I do not know that much about it anyway, but it is quite a story. There will be a lot to learn from it.

Let me be clear: There are dozens of Solyndras out there. Dozens. Many of them took large amounts of government money. But I think the Solyndra story is notorious because the company was really good at promotion. They were able to entice Obama to come out. When they built a new building, he was at the dedication; there was a lot of speechmaking and so on.

In the end, we got what we deserved. Mistakes are often about people; they do not just happen. There has to be somebody turning the dials to make a mistake. I am not going to say I was not responsible. I was an investor and board member of the company, and I have to take responsibility. There were mistakes made, but there were also forces that we could not control very well.

After what you've learned from Solyndra, what sort of cleantech investment interests you now?

Let me give you an example.

There is a company called Heliotrope, which came out of Lawrence Berkeley National Labs. They have come up with a coating for window glass that can modulate both the amount of light and the amount of heat that comes into a room. The inventor is a young woman, a professor in their nanotechnology group.

Nobody has ever been able to do that in combination. In addition, the nanotech process they use is much cheaper than the process used today. The math is pretty simple. We can provide a much lower cost of energy and still sell the glass at reasonable profit. Its early-stage focus is a capital-light model because we do not want to find ourselves with $100 million invested in a company only to discover it has no market or that builders will not use it.

Anything you sell to the building market is tough, because there are many people in the channel and they do not always make rational decisions. You have to be really careful and do a lot of partnering. That is why we will keep the investment down.

I am very excited, because it is pure materials science being directed toward efficiency; it does not require you to change the system much. I see a lot of things they can do with software now to control lighting, heating, cooling, and so forth in a building. I think these things can be installed and operated for low cost.

157 The Solyndra bankruptcy and shutdown resulted in widespread criticism of the U.S. Department of Energy's loan guarantee program. See U.S. House of Representatives, Energy and Commerce Committee, "Committee Releases Extensive Report Detailing Findings of Solyndra Saga," 2 August 2012, https://energycommerce.house.gov/news/committee-releases-extensive-report-detailing-findings-solyndra-saga/.

So, what does Solyndra mean for cleantech today and going forward?

Between the Chinese underpricing and the radical overinvestment, it was like the tide was in and out simultaneously. Solyndra is not the only ship that sank. There are still U.S. companies going out of business today. I read about a company last week called HelioVolt that NEA was involved with, along with several other investors. I don't know how much money was involved, but there was a lot of money and now they're out of business.

Yet the industry itself is starting to boom now. Some of the innovation is really compelling, and a lot of it is occurring outside of the United States. The U.S. industry is not investing a penny in R&D; Asia continues to invest very heavily. Efficiencies keep going up. Just last week I heard some efficiency numbers that I would have never believed to be achievable.

These things fundamentally—especially in the energy area—are about manufacturing cost curves and prices to see demand. You have to get them harmonious with each other. You can keep reducing your cost and your selling price. You can sell more. It's really simple.

Demand is fundamentally driven by prices, while cost is driven by quantity. If you have a manufacturing cost curve, your price-demand curve has to always be above the manufacturing cost curve. That is profit. In solar, the price-demand curve has really been all over the place for the last 10 years.

More recently, you've been a partner at the venture capital fund Formation 8. Can you talk about that?

I helped Formation 8 raise their fund and also served as a mentor to the partners. We struggled to raise that first fund in that it took about a year and a half, but we ended up raising $450 million—that's probably more than anyone has raised for a new fund since 2000.

I only intended to stay at Formation 8 long enough to see some successes, and then go back to my family fund, Baruch Future Ventures. I ended up staying at Formation 8 while working on Baruch Future Ventures in parallel however.

As of today, the returns in the fund are incredibly strong—we've really knocked the ball out of the park. The returns in Formation 8 Fund I are over 100% IRR after three years.

A 100% IRR for a fund as large as $450 million is incredible. What's the hypothesis and investment thesis for your personal family fund, Baruch Future Ventures?

I created a hypothesis that the world is running out of resources. The biggest societal problem is that population is growing sufficiently at a rate of about 2.5% a year such that there will be probably be 11 to 12 billion people on this planet in 2050, and currently there's about 7 billion. How are we going to feed those people, and where are they going to live? What are they going to do for energy? What are they going to do for clean water?

And yet at the same time, we have superb technology available to us now—not only as specific silos of technology, but also arising from the convergence of different technologies. I call it the multidisciplinary technology approach to solving problems.

The technology sectors that I wanted to address are in energy, water treatment, air quality, food security, and healthcare therapeutics. I try to identify emerging seed deals that have the potential to be transformative, and then apply everything I've learned over the years to help those seed companies mature and become large transformative companies.

In April 2015, you sold Taxon to DuPont at a step up of about 8X after 18 months. This is one of your investments at Formation 8. Can you talk about that?

Taxon was a company founded about 12 years ago by a couple of really visionary scientists, and they chose to work on characterizing different microorganisms found from seabeds on the ocean floor. They got a lot of samples from Chevron, which does a lot of offshore oil and gas production. In characterizing the microorganisms in these samples, the scientists found unique combinations of microorganisms. They were looking for organisms that could eat very large molecules called *asphaltenes*, which are high-carbon content compounds.

Chevron supported that research over the years—not a huge amount of financial support, but enough to keep the lights on. When we started looking at Taxon, I thought, "Why are we doing this?" I had actually worked with asphaltenes when I was at Exxon, and I know they're very difficult materials to work with.

Then I realized that there was valuable IP developed at Taxon, but the they were using it for the wrong application. Because I'm interested in food security, it occurred to us that if we could apply these consortia of microorganisms to facilitating agriculture, we might have a very useful product.

We ran a trial with corn, and plants that had been doused in our microorganism were about 25 to 35% more productive, and taller with heavier leaves. These microorganisms were able to uniquely facilitate the transfer of nitrogen and phosphorous from the soil, so the plants just were more efficient in doing their thing: growing.

We were working on how to commercialize agricultural asphaltanes and were talking to a number of potential partners. DuPont ended up purchasing Taxon at a nice price. We had only been investors in the company for 18 months, and it was a really good IRR.

Looking back, that good return started with identifying a problem. One huge problem going forward is making enough food available for 12 billion people who are going to be hungry three times a day.

Can you talk about your investment in Calysta?

Calysta is a company that started pursuing a philosophy of "think big, start small," focusing on making various products from natural gas. Calysta started with a goal that many have pursued over the past 50 years—being able to convert gas to liquids—which can be useful for creating energy or making things. Synthetic biology is a rapidly emerging field that's being enabled by things like low-cost sequencing and computational biology, and Calysta had a strong team of synthetic biologists.

Initially, we pursued a model very similar to what I did at Symyx, which is to create a platform technology, adapt it to specific customer needs in different vertical markets, and then license it to different people. Then, with the price of oil declining rapidly, we realized that oil was still going to be a feedstock for doing a lot of things, such as making plastics, which reduced the market for our technology. In addition, I became more disaffected with the platform model. I came to realize that it's more important to go vertical than horizontal—vertical stretch versus horizontal spread.

We needed to pursue a specific product or opportunity, get as far into the channel as we could, and influence the ability of a customer to solve a pain point. As a result, we ended up identifying that there's a huge problem in aquafarming.

About 50% of the seafood we eat today is raised on aquafarms, not fished out of the ocean. Ironically, those fish raised on aquafarms are being fed fish that people are harvesting from the ocean, so you're not really solving anything from an environmental perspective. In fact, it takes two to six pounds of ocean-harvested fish to feed one pound of aquafarm-grown fish. The numbers are really bad. For example, the fish used as feed are harvested off the cost of South America, and then they're shipping that to Norway and Sweden where all the big aquafarms are—so in addition to the folly of feeding fish to fish, you're using a lot of energy and so forth.

We started trying to get a better understanding of the supply chain and who were the players, and identified a big problem. Fish food had increased in price for the last 25 years at a rate of about 7% a year. Meanwhile, our natural gas had declined in price by almost an order of magnitude. We began to investigate how we could apply our synthetic biology technology to make feed from natural gas; basically, we could use very cheap natural gas to make what is called single-cell protein, which we would then feed to aquafarm fish.

This to me is a quintessential project—and exactly the kind of thing I want to do—where a company solves a tough problem with an economically rational solution. You can apply technology to solve a serious problem that benefits mankind, and at the same time make a good profit for your investors.

You mentioned earlier that the U.S. industry is not currently investing in research, which hampers the development of innovations that improve energy efficiency. Are there workarounds for the lack of investment?

Venture investing is not always the best avenue. Given the mathematics of compound interest, there is no market big enough to justify a 20-year wait to pay out.

My wife and I have funded a center at Rensselaer Polytechnic Institute[158] and we are looking at research on photosynthesis. Photosynthesis is something everybody knows about, we learn about it in high school biology. Yet the fundamental physics is not understood to this day because it's a very complex series of reactions. This group at Rensselaer, which is led by a professor named Lakshmi, is on this trail. They are determined to figure it out and I hope they do. I would never invest in that as a venture capital project, though, because it could take 20 years.

Good technology can make a big difference. By "good technology" I mean technology that can be leveraged into a product or service that a customer really needs—at a cost that provides a value proposition to the customer while making money for the producer. That is the whole name of the game.

A lot has changed in the first ten years of cleantech, including exponential leaps and bounds in technology and R&D. What facets of cleantech are you excited about now?

One of the fields that I love is research productivity.

As a research productivity investment, I got involved with a man named Alex Zaffaroni, who is well known at Stanford. Alex and Peter Schultz were the first people who started using combinatorial approaches to discover new drugs—they mix and match various targets and various chemicals to try and make a hit. The hit is to find out where there is some interaction that you can measure by a physical phenomenon, whether it is light emission or change in magnetic capacity or whatever.

The combinatorial approach changed the research process for drug discovery from what had been a serial process to a parallel process. It was an analogy of Moore's Law, when Moore's Law went from hierarchical computing to distributed computing. Before I got involved I was saying, "Research, especially in materials science, is very unproductive. It takes 10 years to discover something. Then you go through the applications period, and that could be another 10 years. Who is ever going to make money in this kind of thing?"

For example, we've proven through Symyx Technologies and a number of

158 The Baruch '60 Center for Biochemical Solar Energy Research at Rensselaer Polytechnic Institute is funded by Tom Baruch and his wife. The center is currently looking at research on photosynthesis. See *RPI News*, "Trustee Makes Donation to Start New Solar Energy Research Center at Rensselaer," 31 October 2008, http://news.rpi.edu/luwakkey/2508.

children and grandchildren of Symyx that we can reduce the time for materials discovery from 10 years to one or two years. That's huge.

I am proud of that, and I am still pursuing investments in research productivity, but now it is in the software aspects and also more on the application side. I am doing other things today that address that area using some of the Internet socialization schemes particular to analytics.

Is there anything you'd like to say about the energy sector or energy investing that we didn't ask you about?

I think that energy investing is very difficult if you are trying to create electrons and trying to take on oil and gas. I think it's all right if you find pain points and you can identify where big oil and gas will pay for something they need. It's probably not going to be a huge business, but may be a reasonable, profitable business.

I think the "better, faster, cheaper" area of the energy sector—efficiency, especially in buildings—is a better option.

Insights

Corporations add value as LPs. Traditional venture capital funds typically raise capital from multiple institutional investors, foundations, endowments, and high net-worth individuals. In contrast, corporate venture capital funds, like Intel Capital and Google Ventures, have all of their capital from a single source: their parent company. For cleantech, a mixed approach may be best, as Baruch took with CMEA Capital in 1989: he raised from multiple corporations, notably 3M and Dow Chemical, and leveraged their expertise to make investment decisions. For venture capitalists, strategic partnerships with large corporations as LPs allow a fund to both expand its deal flow and benefit from its LPs' due diligence capabilities in evaluating new technology. It's a win for the LPs as well: not only are they exposed to technology and applications outside of their own companies, but they can also make follow-on investments in companies that align with their own business capabilities.

Small funds should avoid cleantech. It is important for a fund's timeline to match its size—small funds have difficulty investing in cleantech and biotech companies because they lack the capital to stretch out over the long development cycle. Small funds should be very wary of cleantech, as they probably won't have the opportunity for an exit within their fund's time frame.

Trying to make energy cheaper is a losing battle for solar startups. The energy industry has historically tried to solve one problem: the cost of energy. The mandate has been, "Make energy cheaper!" But this demand for lower costs has, ironically, been both an obstacle and a pitfall for the industry. Solar companies in the United States had been successful at bringing down manufacturing costs—until they were met with unprecedented low prices from China. Companies like Solyndra were then forced to sell their products below cost, eventually pulling themselves into bankruptcy. Today, it is generally agreed that manufacturing is no longer the path to profit in the solar industry; solar companies are instead finding success in business models designed around the services and financing aspects of the industry. Early-stage investors are wise to shy away from manufacturing companies and instead look to finance and services companies in the solar sector.

Social impact can be profitable. Social entrepreneurship is more than just the ethical "right" thing—approached from the perspective of solving the quantifiable problems of the future, it can also be very profitable. For example, Baruch realized a need to address the long-term limitations in food supply, and was able to pivot the application of Taxon's technology to significantly boost the yield and crop quality of corn. DuPont's acquisition of Taxon contributed to the 100% IRR for Formation 8. This same framework helped Baruch and the company Calysta identify problems with aquafarming, the process that creates 50% of our seafood. Upon inspection, Calysta discovered the utterly backward economics of seafood development, and saw their opportunity. After applying their technology, Calysta helped aquafarmers drive down operational costs, create a business model that efficiently uses resources, and sustainably meet market demand. This notion of profitability from social impact is the premise of Baruch's new personal venture fund, Baruch Future Ventures.

Artur Runge-Metzger

Director of International and Climate Strategy, European Commission

Artur Runge-Metzger is the only international climate policy negotiator in our set of interviews; additionally, he is the only one who holds a PhD in agricultural economics (or any other type of economics), which is unusual for someone in a negotiator's role. In 1985, he joined the faculty at the University of Göttingen in Germany, teaching about—and researching on—rural energy and agricultural development.

In 1993, Runge-Metzger left Göttingen to help run the European Commission's cooperative program on agricultural, rural, and environmental policies in Zimbabwe. He moved to Brussels in 1997 to work in the European Commission's Division of Development and Environment until 2001. After a few years on a mission for the EC in Bosnia and Herzegovina, he returned to Brussels to become director of the EC's Negotiation Unit on Climate Action.

In 2009, Runge-Metzger assumed his current position as the European Commission's Director of International and Climate Strategy, Climate Action, making him one the most influential climate policymakers in the world. Further indication of his influence can be seen in his nomination to co-chair the United Nation's climate talks from 2013 to 2014. Over the course of his career, Runge-Metzger has emphasized the development and market introduction of innovative technologies to meet the emerging challenges in both the agricultural and energy development sectors.

Do you think the policy instruments employed so far, including the ones being used in the European Union, have been as economically efficient and equitable as you personally would like to see? How do you think about those two policy dimensions?

Support has been given in excess of what would have been necessary. A number of policies are examples of what economists, who really strive for efficiency, would call "gold-plated." Those would-be subsidies for clean technologies that stimulate technological development and the early introduction into the marketplace of these technologies. This subsidization strategy works very well in the early days of new technologies, in terms of jumpstarting those technologies down their learning curves toward lower costs.

These subsidization policies really drive renewables into the system, but only up to a point—when there is enough volume that financing the programs is going to hit you. . . . I think this is what we have seen over the last two or three years in Europe. A number of member states had to reverse their policies of large subsidies for renewables, for instance, depending on exactly how they had designed them.

These policies have been designed in very different ways. If you look at Spain, the stimulation of renewable energy technologies was done in a way that affected the public budget: it was a support and a subsidy that would have to come out of the public budget. So, as soon as they went into a budget deficit and had their financial crisis, it was completely impossible to maintain. In Germany, it is the consumer who is paying the bill for driving renewables into the system, with guaranteed feed-in tariffs.[159] That has caused an uproar in small or medium-sized industries there, but now even residential consumers have said, "No, this is getting too expensive."

In that respect, these policies were effective, but they were not efficient. A lot of changes had to be made, which reduced what was paid for renewable energy supplies—and upset the investors in those renewable energy projects, who had hoped for a clear stream of income heading their way in the coming years. Many of them have been quite frustrated over the last few years. As a result, we saw a drop in investment in renewable energy.

I think you can say the same about the transport sector. Reducing CO_2 emissions from cars via engine modifications is not cheap. The only thing that has helped reduce CO_2 emissions from cars and trucks is that gasoline prices have basically been rising steadily for more than ten years now. In addition, there is very high taxation on gasoline in Europe, which would equal an enormous carbon price if you translated it that way. You'd be talking about 100, 200, 300 Euros per ton of carbon dioxide.

Any other comments on this point?

In the European context, as you can imagine, this tradeoff between short-run efficiency, longer-term new technology stimulation, and equity is also a very important point. Our policies. . . if you look at any of the varied interests here, when heads of state come together, they always talk about efficiency and fairness

[159] Electric utilities use *feed-in tariffs* to pay for renewable energy supplies at cost and roll the additional expense into the average rate charged to all their electricity consumers. While this type of policy does encourage expanded use of all types of renewable energy, it can result in unequal costs among renewable energy sources and higher electricity rates if the cost of renewables is higher than that of conventional energy sources. See glossary for additional details.

in the same sentence. This happens because we have poorer member states in the European Union, where there is still a lot of mitigation potential, and we do our international climate negotiations collectively.

From an EU-wide efficiency perspective only, these poorer nations would have to carry the brunt of the investment if only the cheapest mitigation options are to be undertaken, and that's not necessarily possible for them. This problem is always very conflictual, and it is one that needs to be resolved. I think there are ways to try resolving issues of equity, fairness, and efficiency at the same time—even if you use market-based instruments.

With a market system, no matter how you allocate the burden of CO_2 emission reductions, you will always arrive at an efficient outcome. That efficiency might be a little more difficult to achieve strictly with taxation, because then one would have to think of differentiated taxation and so on, to protect the poorest citizens from bearing the highest costs. You then move away from getting to the efficient solution at the end of the day.

In an emissions-trading system under a cap on total emissions, efficiency is possible, but damned difficult to fully achieve: in making efficiency versus equity tradeoffs, you have to make assumptions about the future of carbon prices. If these assumptions fall flat, your redistribution and your fairness angle might also fall flat.

I often think that in Europe we have learned quite a number of lessons about how to make these difficult tradeoffs. As you say, we have to improve when we do the next iteration.

From your perspective—and you may be one of the best people in the world for this question—what role do you think international cooperation should play in promoting the energy transformations the world needs?

International cooperation is very important from different perspectives. It is important on both sides, whether research and technology development or creating the right policy environment in your country in order to drive innovation.

There are many areas where we try to work together with other countries. In developing countries, very often the focus is on technology development. Then the expectation is that transfers of technologies from developed countries can fall like manna from heaven, which of course is not possible. There are difficult IP issues and capacity issues that must be addressed.

We have been focusing more on improvements in the policy environment and the enabling environment—to get those right—and to have more technical cooperation on both sides. I think that is paying off quite well. There are many developing countries, for instance, trying to create the right environment for renewable energy introduction.

China is trying an emissions trading system. We in the EU have been working a lot with them over the last few years to put a price on carbon—that could drive the whole thing. That is an important aspect for us.

When it comes to cooperation between Europe and the United States, I think there is an awful lot we can do, particularly with developing certain testing standards. At the moment, for instance, when the United States adopts testing standards for heavy-duty vehicles, it would be good to have the same testing cycles in Europe, so we can compare.

The corporate developers who create products for a much larger market—not just for Europe and the United States—will help drive down the cost of technology a lot. I think technology cooperation is very important.

Some people have characterized what's needed at the highest level for the development of effective climate policy as new technologies and institutional change. How well suited do you think the current set of institutions around the world are to achieving our objectives? Do you have any personal, favorite recommendations for new institutions, or can you suggest major reforms for existing institutions?

I'm always somebody who thinks it's better to work with existing institutions than to try and see if there are any gaps that might require establishing new ones. That way, it's the existing institutions that pick up any new responsibilities.

We have had some institutional innovations in the past, like IRENA [International Renewable Energy Agency], working on renewable energies. I think they are doing a good job. I think that the International Energy Agency [IEA] has been very helpful over the last few years; it has become more focused on trying to strongly integrate climate into their policy agenda.[160]

That climate policy focus, I think, has also been helpful in the IEA's technology perspective work and so on. We shall see whether those institutions created under the United Nations—the Climate Technology Centre and Network—are going to work. There is huge potential there, but that needs to be grasped.

Then there are many very specific initiatives. I think the establishment of the Global CCS Institute in 2009 was a good thing. It works out of Melbourne, Australia, on carbon capture and storage, and so on.

I think it's also helpful to have good cooperation with industry in very focused initiatives trying to resolve particular topics—that is also key. At the end of the day, a lot of the work needs to be done by governments and private sectors working together.

In your opinion, are there any big clean-energy technological challenges not currently receiving adequate attention?

I think one should probably not put all the weight for needed innovations on

[160] See International Energy Agency, *WEO 2015 Special Report.*

these existing institutions. One of the big questions that still needs to be resolved, certainly, is the whole question around energy storage. That is an area where I think I would put a lot of that particular money.

Another area is around carbon capture and use technology, particularly those aspects that look at whether we can make any use of the CO_2 stream in manufacturing high volume products. There's a lot of little research projects happening here and there, but it has really not gotten up to scale.

CO_2 emissions will be with us for a very long time, so we need to find a solution to that. I think now is the right time to start research in that area. Energy storage and CO_2 emissions are two areas I would probably put high up on my list for more attention.

Would you add additional work on so-called smart grid technologies to your list?

That area, certainly, in terms of the whole management of energy to make it really smart. I think there is a huge potential that has not really been tapped yet. I'm sure with IT there will be much better solutions in the future.

Are there any things in the EU climate policy space that you're particularly proud of or not-so-proud of—things that you wish you had not done, or done in a different way?

What I'm particularly proud of, of course, is that we were able to start the European Union Emissions Trading System [ETS], even if we had to make a lot of compromises in the beginning to get it established. There was a big national hand in the initial design of the Emissions Trading System.

I think what we still want to address now are these interactions with other policies, so that the next financial crisis or the next big implementation of a renewables system or an energy efficiency revolution will not undermine the workings of the ETS.[161] I would say that coordination problem is something we have overlooked in the past. Then when it comes to the other policies, like renewables, I think the energy sector needs to try to get to a level playing field so that all energy-

[161] It is well documented that, in its initial stages, the price of an emissions permit within the ETS collapsed down to near zero because the amount of energy efficiency and renewables required by the EU nations led to an oversupply of carbon emission permits. See, for example, A. Denny Ellerman et al., *Pricing Carbon: The European Union Emissions Trading Scheme* (Cambridge, UK: Cambridge University Press, 2010); and Markus Wråke et al., "What Have We Learnt from the European Union's Emissions Trading System?" *Ambio* 41 suppl. 1 (2012): 12–22, doi:10.1007/s13280-011-0237-2.

Later, the confluence of high renewables subsidies in some countries and the global financial crisis, as mentioned above, led those countries to cut subsidies for renewables, which in turn led to higher prices for the emissions permits.

supply technologies compete on a least-total-cost basis. That is one of the big things that still needs to be done.

That standardization is one big thing we still need to do on renewables in the EU as well, instead of having a patchwork of policies implemented by the individual members of the European Union, which makes it very difficult for investors, the finance markets, and the big energy companies to do their best.

Changing subjects now, what is your current thinking about shale gas? There's a bigger on-the-ground expansion in the United States than elsewhere, but the shale gas revolution may spread pretty quickly around the world. If it is unwise to go too far in this direction, what kind of policies do you think could be used to help moderate the move to shale gas?

I think the final judgment on shale gas is that we'll have to wait for a couple of years. In terms of the carbon emission effects, it allows the possibility of a fuel switch between coal and gas. It's one of the options we have in order to reduce emissions. That certainly is happening in the United States. It's what we can clearly see.

The relevant question now is more like, "What are the secondary effects in terms of the impacts of shale gas development on other fuel markets?" Coal is getting cheaper as a result. At least in Europe, where we face quite high gas prices, that has led to a kind of coal renaissance. Of course, that was only possible because the Emissions Trading System has created more emissions credits than will be necessary to reach our goals, which will go on for quite some time. So far, the ETS hasn't put a lid on that kind of coal development.

I think one of the other questions is, "How much do you really save in terms of future GHG [greenhouse gas, a.k.a. carbon] emissions with shale gas?" I think there are still some unknowns concerning methane leakage from the gas system. How much is it, really? That is going to be different from country to country, depending on how good the drilling is and how one can prevent fugitive emissions during that process. I think it's still not clear where this story is going to end—it will very much depend on each geological situation where drilling will take place.

In 2014, our understanding was that the 40% reduction in GHG emissions relative to 1990 levels by 2030 target was maintained by the United Nations Framework Convention on Climate Change (UNFCCC),[162] but then there was some loosening of renewables program targets in different countries

[162] One of us—John Weyant—was privileged to be at the AMPERE stakeholder meeting right before the ministers' meeting in Brussels in January of 2014 where the tentative plan for the EU for the next United Nations Framework Convention on Climate Change (UNFCCC) budget period was announced. He got into the meeting the day before the ministers met and made the announcement, and there was a lot of talk about easing up on overall targets.

for exactly the reason you already mentioned—the relative cost burden and equity between EU members. Are you pleased with the direction that has taken? Would you have liked to have seen it go further or be tougher? What's your overall assessment?

Of course, we expected these discussions would have been finalized in March of 2014, but that was delayed.

What really got in the way of this agreement is this whole question of fairness. What does it mean for each of the member states, at the end of the day, to adopt and accept the 40% target? There are certainly some poorer member states who say, "You all talk about reducing emissions. If you look at the most-efficient least-total-EU-cost case, then a lot of that will have to be done in the poorest countries in Europe. You richer countries are calling for it, but we will have to pay the bill."

That question had to be politically resolved at another discussion of the European heads of state. In terms of comparing the structure of the new package to what we are doing right now—we're reducing the renewables and loosening the energy efficiency requirements in each country. We never had a very rigid system in Europe.

I think the leaders of the EU are pleased the negotiations moved in that direction because at the end of the day, we think that only the Emissions Trading System will guarantee these things at the least cost. We are not going to get there if we spend too much money on things that are too expensive. We need to look at and put more emphasis on the efficiency of the system in terms of achieving our emissions targets at least cost.

AMPERE had one distinctive work package[163] that considered cases where the European Union acted alone and nobody followed, and ones in which the EU acted and people followed without too much delay. It was noteworthy and relevant that they did try to look at the first-mover advantage explicitly in this clean technology space, which was an outstanding contribution to the analysis side. Did that have any traction in political circles?

Yes, this argument still has some traction. I think there are some good examples where we can show that, in terms of real numbers and figures, the prospect of this kind of first mover advantage really helps. If you look, for instance, at vehicle

[163] The AMPERE work package referenced here was primarily proposed by Pantelis Capros's group with the National Technical University of Athens. See Elmar Kriegler et al., "Making or Breaking Climate Targets: The AMPERE Study on Staged Accession Scenarios for Climate Policy," *Technological Forecasting and Social Change*, doi:10.1016/j.techfore.2013.09.021, https://www.pik-potsdam.de/research/climate-impacts-and-vulnerabilities/projects/project-pages/world-bank-report/publications/kriegler-ampere-staged-accession-tfsc14.pdf.

efficiency standards and all the technologies around the automobile engine, certainly the European automotive industry is at the forefront, as it has the largest share of patents in that area.

These improved energy efficiency engine technology innovations have definitely been driven by having efficiency standards on engines and tightening them rather quickly over time. It's certainly an advantage. The same probably is true for renewables: if you look at the development of wind and so on, a lot of that happened in Europe, giving a first-mover advantage to the technology companies here.

Insights

Policies need to consider effectiveness, equity, and efficiency. Spain and Germany's feed-in-tariffs, for example, were effective but not efficient and therefore were not sustainable in the long term. Creating a market-based system for CO_2 emissions reductions could be efficient but not fair; lower-income, higher-polluting countries would potentially bear disproportionate costs, which ultimately would reduce the effectiveness of the policy as these countries opt out.

Energy innovation policies should be flexible. The overall carbon emissions reduction program in Europe is quite complex. Despite some growing pains—such as very low emissions-permit prices and high costs of renewable standards in some countries—the system has had some flexibility built into it, and is being adjusted for the 2020-2030 GHG emissions budget period. The initial Emissions Trading System in the EU included trading of emissions across national boundaries only in the industrial and electric utility sectors, whereas individual countries set goals for energy efficiency and renewable energy. In the next round, it will be important to make the energy efficiency and renewable policies more consistent across countries, and to move their costs closer to the price of an ETS emissions permit. These adjustments will help lower the overall cost of the emissions reduction program.

Looking Ahead: The Interdependence of Industry, Policy, and Finance

The world will not go carbon-neutral in the next decade, but that does not mean failure for the energy industry. To the contrary, a wide array of compelling initiatives and projects is just getting off the ground. We believe the next cycle can be more efficient and predictable than the earlier ones, with industry's adoption of the lessons of the past decade.

So what is the future of clean energy?

If we summarized this publication into one point, it would be the importance of considering every clean energy initiative as a stool needing three legs for stability, predictability, and mutual reinforcement. In cleantech, those legs are industry, policy, and finance. As we reflect on the biggest insights from the first decade of cleantech, as shared by our interviewees, this interdependence comes to the fore.

First and foremost, industry experience matters. Cleantech startups have been operating at a disadvantage. The traditional energy industry has not fundamentally changed in at least 30 years, which means that all its processes are well-tuned through time and experience. Every aspect of an oil company, for instance, runs with tremendous efficiency based on tried-and-true employee experts who know what they are doing. Contrast that with a clean energy startup: although many of the founders and employees may have deep experience in a particular aspect of their enterprise—say, materials science or software—they often utterly lack experience in other areas, such as manufacturing, infrastructure, business development, or operations.

This asymmetry is not much different than what one might find in traditional IT startups, but it is a much deeper problem for clean energy. Many IT startups are working within relatively new markets with new technology and can be allowed a relatively sloppy execution or inexperience with business models. Clean energy companies, in contrast, are creating new energy markets with new technology while being forced to compete against the existing energy market and juggernauts that are mature and experienced in all aspects of their business, not just its technology.

The good news is that there's an increasing crossover of expertise between existing energy and clean energy. Moreover, there is now a whole crop of weathered experts: those veterans who survived the first clean energy cycle. Future and ongoing endeavors in the space will be much less disorganized, and better at getting the basics right the first time.

Policy affects capital availability for clean energy. If one thing could be significantly changed about the last capital market cycle, it should be the stability and diversity of sources of finance for clean energy companies. The word *cycle* sums up the entire problem: Access to capital is 99% governed by large, exogenous forces, and 1% governed by a company's timetable.

The lack of diversity in clean energy finance led to an entire generation of companies being wiped out, merged, shut down, or sold. Policy is simply that important—it can have a dramatic effect on factors including technology readiness, finance, and marketing, which are needed to foster the clean energy cycle all the way from conception to market penetration.

Appropriate government policies have a large role to play in energy innovation. While past government policy has helped drive a boom-and-bust cycle, it has also assisted in the creation of real innovation and led to billions of dollars of private investment. A monoculture policy around energy innovation won't suffice, but a careful mix of heterogeneous policies can promote a much healthier ecosystem without the previous, lopsided outcomes.

The roles of the public sector (policy) and private sector (industry) are inextricably intertwined. The most important element in terms of policy is the centrality of private-sector collaboration in driving that appropriate mix, in both creating policies that educate consumers about current and upcoming energy alternatives and facilitating access to the widest possible array of these technologies and alternatives.

One key government role is to support advanced technology R&D that is too speculative for the private sector. In fact, such advanced technology IP is difficult to define, much less protect, before commerce-driven firms develop their own versions of the commercial technological results. The government should not be in the business of picking winners and losers among competing product ideas; the private sector is much better at doing that on a risk-adjusted basis. On the other hand, government can—and should—sponsor continual improvements in the research-driven technology base from which the private sector draws for new businesses, attracting the leading researchers at national laboratories and universities. The private sector should not take on more technological risk than it could normally justify, any more than the public sector should make individual venture investments.

The two sides, public policy and private sector, play distinctly different roles in the innovation process that are strongly complementary. Good communication across the sectors is important, but role reversal can be—and has been—calamitous.

A second key government role is to inform the private sector about a particular technology's readiness. Companies can then avoid trying to build a business plan, financing approach, and marketing strategy whose core cost exceeds the available money in the current or foreseeable marketplace. If the private sector pursues a technology before it is ready (either accidently or intentionally), it needs to simply let that project go. Government attempts to implement policies to fix such a mismatch can get expensive, and should be avoided.

Once a technology gets close to market competitiveness, there are then a whole host of additional policies in financing, technology demonstration, and consumer information that can be vital in getting the technology through the "Valley of Death." There's the barrier of market entry cost that new technologies face as they need to scale up production in order to compete economically. Thus, the best role for government is to provide more technology concepts for those entering the Valley of Death to choose from, and to help pull those companies that meet the technological feasibility and cost-competiveness criteria into the marketplace.

In this area, cleantech's past has provided two main lessons. The private sector should not pursue new ventures before the technologies involved are ready for commercialization; conversely, the public sector should not try to make new venture-style investments. Government should not provide favorable financing to early-stage ventures for which there is still significant technical uncertainty, but favorable public financing should be available to ventures with minimal outstanding technical uncertainty and close-to-market costs.

Cooperation is absolutely essential. Ultimately, clean energy policies are meant to drive technology, which exists on a spectrum from energy-supply technologies to energy-demand technologies, with a number of gray areas between the two. One extremely important example of an in-between technology is electricity-grid integration technologies. These hardware and software systems can greatly increase the amount of renewable electricity generation that can be added to electricity grids at low cost and with minimal disruption.

This past market cycle has proven the need for equal, measured advancements for both supply *and* demand. These advancements have previously appeared random. The guiding hands of policy and finance are particularly vital given the fiendishly cyclical nature of the industry. It is imperative to explore new ways of financing energy innovation through policy initiatives, and not just at the national level—state and local policies can help a great deal too. The good news is that change has been happening at the state level, in places like California, New York, and Texas.

The word *cooperation* cannot be overused in the context of new policy initiatives. Cooperation is needed not just among national, state, and local authorities, but energy innovation policies must also reach across other infrastructural areas, such as IT, national security, water, food, and health. In a world where industry treats energy as a commodity market (in which the cheapest and most scalable always wins), the contrast between the efficiency of a commodity market and the evolution of careful policy could not be greater. The biggest insight here is the need to take much greater care in balancing a policy's efficiency against equity and implementability during the design phase. Ultimately, this cooperation will need to be institutionalized, alongside innovation in support of energy policy.

Investors must balance risk and security with a portfolio approach. When people hear the word *finance*, they often think of numbers, spreadsheets, math, and a host of extremely conservative folks in charge. That image could not be further from the truth for clean-energy venture capital. Consider the difference between financing a shipbuilding enterprise and financing a single intrepid explorer's trip across an unknown ocean. All aspects of the shipbuilding business are known, so investment in shipbuilding can provide predictable returns. An exploration is full of unknowns as well as known risks (e.g., hurricanes) that might wipe out the investment completely; however, those who do return with new, lucrative trading connections will win big, so the risks may be acceptable.

This balance of safe investment and risky-but-potentially-lucrative investment is the core of venture capital, and cleantech financing is certainly not an exception. A portfolio approach promotes some certainty, making it acceptable for investors to lose on some ventures because others in the portfolio will pay off.

Investors need government to create markets. Taking the shipbuilding enterprise analogy further, governments have a critical role in developing burgeoning trade routes and trade partners. This role includes supporting shipbuilders and individual explorers as well as providing access to additional capital and markets. When cleantech previously encountered major difficulties, policy and government support was lacking, which contributed to the loss of an entire generation of clean energy companies. The good news is that there will always be interest in exploration, and the winners in clean energy will win big.

A new wave of cleantech is underway, and this time the cleantech industry as a whole has insights and experience—for instance, these interviews indicate that the proper role of governments is to create markets. Instead of funneling massive investment into a single company, a government should find ways to build programs that drive demand and allow companies to find scalable business models early on.

Carbon markets are one example. By creating a market around carbon emissions, early-stage clean energy companies backed solely by venture capital are able to start making sales and revenue right away, which largely defrays dependence on significant outside capital at scale. In fact, there are many ways government can create a friendly ecosystem for venture-backed companies, from regulatory approaches that encourage multi-tenant building owners to "go green," to tax credits for consumer purchases of energy-efficient vehicles.

Government needs investors to identify talent and place bold bets. People and personalities, not technologies, drive change. Venture capital can provide this irrefutable aspect of clean energy—by finding the bold entrepreneurs of the world and making a big bet on their ingenuity before it's known whether the markets or technologies will suffice. Every venture capitalist we interviewed consistently supported one theme: the practice of how they evaluate entrepreneurs and deploy investment capital is based on intuition and pattern matching. No government policy can replicate or improve this process directly, which is both good and bad news: venture capitalists are what we have to work with for the next cycle.

Overall, the first wave of cleantech provided many insights to learn from. Everyone made mistakes, and took the blame for them. We are hopeful that the industry will be much more successful in the new wave of cleantech, because we have not only learned from the first cycle and know what to do better individually, but also because government, industry, and finance will collaborate to support one another.

Whether there will be a cleantech revolution, only time will tell. What we do know is that the scale of the required transformation, along with the complexity and uncertainties of the required systems, is still large enough to challenge the world's best thinkers for the coming years. We hope readers will join us in pioneering and shepherding cleantech's evolution.

About the Authors

John Weyant is a professor of management science and engineering at Stanford University. Honored as a major contributor to the Nobel Peace prize awarded to the Intergovernmental Panel on Climate Change (IPCC) in 2007, Weyant has been a lead author for the IPCC for chapters on integrated assessment, greenhouse gas mitigation, integrated climate impacts, and sustainable development. He most recently served as a review editor for the climate change mitigation working group of the IPCC's fourth assessment report.

Weyant's research focuses on global climate change policy systems and analysis, energy efficiency analysis, energy technology assessment, and models for strategic planning. He is the Director of the Energy Modeling Forum (EMF) and Deputy Director of the Precourt Institute for Energy Efficiency at Stanford. He is also a Senior Fellow of the Precourt Institute for Energy and an affiliated faculty member of the Stanford School of Earth, Environment, and Energy Sciences; the Woods Institute for the Environment; and the Freeman-Spogli Institute for International Studies at Stanford. In addition, Weyant was a founder and serves as chairman of the Integrated Assessment Modeling Consortium (IAMC), a five-year-old collaboratory with 53 member institutions around the world.

Weyant has been active in the U.S. debate on climate change policy through the Department of State, the Department of Energy, and the Environmental Protection Agency. In California, he is a member of the California Air Resources Board's Economic and Technology Advancement Advisory Committee (ETAAC) which is charged with making recommendations for technology policies to help implement AB 32, The Global Warming Solutions Act of 2006.

Weyant completed his BS and MS in aeronautical engineering and astronautics and additional MS degrees in engineering management and in operations research and statistics, all from Rensselaer Polytechnic Institute. He completed his PhD in management science with minors in economics, operations research, and organization theory from University of California, Berkeley.

Ernestine Fu is a venture investor at Alsop Louie Partners, a firm focused on cybersecurity, big data, and hard-science companies. She led her first investment within two months of joining the firm and was recognized by several media outlets for bringing a fresh face to venture capital as a young Asian-American woman. Fu is a Kauffman Fellow, Class 17, and is the youngest person to complete the prestigious venture capital education program.

Fu's energy research includes co-authoring *The State Clean Energy Cookbook*, a report led by former U.S. Senator Jeff Bingaman and George Shultz, former Secretary of State and Secretary of the Treasury. The report analyzes and recommends state-level policies for energy efficiency and renewable energy. As a researcher at Stanford University, Fu studied global climate change and its impact on extreme events such as hurricanes, storm surges, and floods. She also designed and taught courses on energy, entrepreneurship, and government.

Committed to public service, Fu authored the book *Civic Work Civic Lessons* with former Stanford Law School Dean Thomas Ehrlich, discussing how and why people of all ages should engage in public service. After starting the nonprofit Visual Arts and Music for Society, Fu served on a corporate philanthropy board for State Farm Insurance. She is also an outspoken advocate for patent reform, immigration reform, and R&D funding.

Fu completed her BS and MS degrees in engineering at Stanford University, with Tau Beta Pi and Phi Beta Kappa honors. She was awarded the David M. Kennedy Prize for the top thesis in engineering and natural sciences, and the Schmidt-MacArthur fellowship for developing circular economy principles. At Stanford, she served as a student representative on the Board of Trustees and a member of the School of Engineering Dean Search Committee.

Justin Bowersock is the founder and principal of Rockhill Advisors, an independent fiduciary that provides tailored wealth management and financial planning services to high net-worth clients across the United States. With 17 years of experience in the energy sector, Bowersock has worked with closely with multiple energy companies and utilities.

Bowersock was a Research Director at the Steyer-Taylor Center for Energy, Policy, and Finance at Stanford University. His research work includes leading a project funded by the Electric Power Research Institute (EPRI) to help tackle the commercialization funding problem for new electric power generation technologies, developing a proposal to use market competition to set limits on subsidy costs in collaboration with the Department of Energy and private investors, and advising Stanford on its buying strategy for $300 million of wholesale electricity.

Prior to Stanford, Bowersock held several roles in the private sector. Bowersock led business development for wind, solar, biomass, and storage projects in the United States and Canada as Director of a €20 billion Spanish infrastructure firm. In that role, he negotiated deals and established partnerships with multiple utilities and other energy stakeholders. Bowersock also supervised a financial analysis team to cover partnerships, pipelines, independent power producers, and diversified utilities at a major ratings agency. He has also worked at Deloitte and Arthur Andersen.

Bowersock completed his BA in history at Vanderbilt University. He holds the Chartered Financial Analyst designation and is a former instructor for the National Outdoor Leadership School.

Justin Bowersock is the founder and principal of RockHill Advisors, an independent fiduciary that provides tailored wealth management and financial planning services to high net worth clients across the United States. With 17 years of experience in the energy sector, Bowersock has worked closely with multiple energy companies and utilities.

[illegible]

Glossary of Names, Organizations, and Terms

Individuals mentioned in the text of this book are alphabetized here by last name.

A

A123 is a battery and energy storage systems company that was founded in 2001. The company's nanophosphate technology aims to produce batteries with higher power, greater safety, and long life. The firm was the battery supplier to Fisker Automotive, and went bankrupt in 2013; it, and Fisker, were acquired by the Chinese Wanxiang Group, although A123 has since shifted its focus away from the electric car market. See a123systems.com and Daniel Gross, "Not Another Solyndra," Slate, 2 May 2014, http://www.slate.com/articles/business/the_juice/2014/04/the_rebirth_of_a123_systems_not_another_solyndra_after_all.html.

Aclara Biosciences was a microfluidics company that developed and commercialized lab-on-a-chip (LOC) technology for pharmaceutical drug screening and genomics R&D. The company was acquired by Monogram Biosciences in 2004. See Bloomberg, "Company Overview of Aclara Biosciences," 2016, http://www.bloomberg.com/research/stocks/private/snapshot.asp?privcapId=24235.

Advanced Research Projects Agency-Energy (**ARPA-E**) is a collaborative U.S. government agency focused on furthering advanced energy technologies, modeled after the successful Defense Advanced Research Projects Administration (DARPA). See arpa-e.energy.gov.

The **Alliance for Climate Protection**, also known as the **Climate Reality Project**, aims to increase public awareness and action on the issue of climate change. Founded by former Vice President Al Gore, the project's mission is to "catalyze a global solution to the climate crisis by making urgent action a necessity across every level of society." See climaterealityproject.org.

Alphabet Energy uses materials science applications and nanotechnology to create thermoelectric generators, turning waste heat into electricity. See alphabetenergy.com.

The **American Clean Energy and Security Act (ACES) of 2009**, also known as the Waxman–Markey bill (H.R. 2454), would have established a limit (cap) on total greenhouse gas emissions and a system for trading emissions rights from low mitigation cost to high mitigation cost firms. Although the bill was narrowly passed by the House of Representatives, it failed to win Senate approval. See Congress of the United States of America, *H.R. 2454 – American Clean Energy and Security Act of 2009*, https://www.congress.gov/bill/111th-congress/house-bill/2454.

The **American Recovery and Reconstruction Act of 2009** (**ARRA**) was designed to stimulate consumer and business spending by boosting U.S. government spending in order to pull the U.S. economy out of the global recession that began in 2008. The bill was dedicated to "Making supplemental appropriations for job preservation and creation, infrastructure investment, energy efficiency and science, assistance to the unemployed, and State and local fiscal stabilization." It authorized about $800 billion of additional government spending, starting in fiscal year 2009. Congress of the United States of America, *American Recovery and Reconstruction Act of 2009*, H.R. 1, http://www.gpo.gov/fdsys/pkg/BILLS-111hr1enr/pdf/BILLS-111hr1enr.pdf.

The **Appliance and Equipment Standards Program** sets minimum energy efficiency for 50 categories of appliances used in homes and businesses, including refrigerators, dishwashers, and air conditioning units. See U.S. Department of Energy, "About the Appliance and Equipment Standards Program," n.d., para. 1, http://energy.gov/eere/buildings/about-appliance-and-equipment-standards-program.

The EU-sponsored **AMPERE** (**Assessment of Climate Change Mitigation Pathways and Evaluation of the Robustness of Mitigation Cost Estimates**) involved 22 partners from Europe, Asia, and North America from 2011 to 2014. The group evaluated options for reducing carbon emissions, including technology and policy limitations, as well as scientific uncertainty about the effects of possible changes. See European Commission CORDIS, "AMPERE," 29 May 2017, http://cordis.europa.eu/project/rcn/98809_en.html.

ARPA-E. See **Advanced Research Projects Agency-Energy** above.

ARRA. See **American Recovery and Reconstruction Act of 2009** above.

Assessment of Climate Change Mitigation Pathways and Evaluation of the Robustness of Mitigation Cost Estimates. See AMPERE above.

B

The **balance of system** (**BoS**) cost for a solar installation includes all of the up-front costs incurred in addition to the price of the modules themselves. Also referred to as "soft costs," BoS costs include customer acquisition, installation labor, permits and inspections, and connection to the grid or battery storage. See U.S. Department of Energy, "Energy Department Announces $7 Million to Reduce Non-Hardware Costs of Solar Energy Systems," 15 November 2011, http://energy.gov/articles/energy-department-announces-7-million-reduce-non-hardware-costs-solar-energy-systems.

Base load is the minimum amount of energy required by an electric grid. Historically, utility planners layered generating capacity: always-on facilities provided the base load, and additional power generation came online for the "intermediate load" and yet more for "peak load." This idea is intertwined with the increasing use of variably producing renewables: solar cells only generate power when the sun is shining, and wind tends to blow more consistently at night than during the day. However, the concept and definition of base load are in flux, with the industry moving more toward a constant, flexible mix of power generation sources from fossil fuels and renewables combined. See "base load," Glossary, U.S. Energy Information Administration (EIA), https://www.eia.gov/tools/glossary/index.php?id=B; Brigham A. McCown, "Baseload Power Will Keep the Lights On," *Forbes*, 27 July 2017, https://www.forbes.com/sites/brighammccown/2017/07/27/baseload-power-will-keep-the-lights-on/#7dc55e442c5b; Kevin Steinberger & Miles Farmer, "Debunking Three Myths about 'Baseload,'" Natural Resources Defense Council, 10 July 2017, https://www.nrdc.org/experts/kevin-steinberger/debunking-three-myths-about-baseload.

Battelle Memorial Institute is the world's largest nonprofit research and development organization. Founded in 1929, Battelle is focused on applied sciences and technology and manages several national laboratories. See battelle.org.

Bechtel is one of the largest EPC (engineering, construction, and procurement) contractors in the world. A privately-held company, Bechtel builds all types of infrastructure projects, from airports to power plants to stadiums. See bechtel.com/expertise/infrastructure.

In a **bottom-up calculation**, energy savings obtained through the implementation of one specific energy efficiency improvement measure are calculated and then aggregated with energy savings results from other specific measures to assess total energy savings. See Evaluate Energy Savings EU, "Bottom-Up Methods," n.d., para. 2, http://www.evaluate-energy-savings.eu/emeees/en/evaluation_tools/bottom-up.php.

Boulder Ionics is a producer of ionic liquids and ionic liquid-based electrolytes for electrochemical applications. Lithium ion batteries use organic compounds, in contrast to Boulder Ionics, which uses inorganic materials. Ionic materials have low vapor pressure, which allows them to operate safely at higher temperatures. The firm was acquired by CoorsTek Specialty Chemicals in 2014. See coorstek.com.

BrightSource Energy designs, manufactures, and deploys solar thermal technology to utility-scale power plants. The firm was originally named LUZ II, based in Israel. See brightsourceenergy.com.

Bundled energy services refers to a utility combining several services to the end-user, such as "bundling" electricity with energy efficiency monitoring and management. See Tom Kerber, "Bundled Energy Services Poised to Overtake the C&I Energy Management Market," *Greentech Media*, 15 June 2016, https://www.greentechmedia.com/articles/read/bundled-services-poised-to-overtake-the-ci-energy-management-market.

C

C3 Energy (now called C3 IoT) is a smart grid analytics startup founded by successful IT entrepreneur Tom Seibel. Its goal is to provide grid integration tools to electric utility companies. It uses big data, advanced analytics, and machine learning to improve the safety, reliability, and efficiency of power generation and delivery. See C3 Energy, "C3 Energy Redefines Utility Customer Digital Experience with New Release of Customer Analytics Applications" (news release), 28 October 2015, http://www.businesswire.com/news/home/20151028005975/en/C3-Energy-Redefines-Utility-Customer-Digital-Experience; and c3iot.com.

CAFE (Corporate Average Fuel Economy) standards were originally set in 1975 to increase the fuel economy of cars and light trucks. The standards require each auto manufacturer to exceed fuel-efficiency targets on average, across all of the cars they sell. The 2017 CAFE standard is approximately 40 mpg; the Obama administration set the 2025 standard at 55 mpg, but the Trump administration has ordered the EPA to review both the target level and year. See U.S. Department of Transportation, "Corporate Average Fuel Economy (CAFE) Standards," August 2014, http://www.transportation.gov/mission/sustainability/corporate-average-fuel-economy-cafe-standards; and Paul Eisenstein, "Trump Rolls Back Obama-Era Fuel Economy Standards," NBC News, 16 March 2017, https://www.nbcnews.com/business/autos/trump-rolls-back-obama-era-fuel-economy-standards-n734256.

The **California Public Utilities Commission** (**CPUC**) regulates privately owned electric, natural gas, telecommunications, water, railroad, rail transit, and passenger transportation companies. The CPUC's goal is to ensure safe, reliable utility service and infrastructure at reasonable rates, with a commitment to environmental enhancement and a healthy California economy. See cpuc.ca.gov.

CalCEF, the **California Clean Energy Fund**, is a family of nonprofit organizations working together to take clean energy technologies from the innovation to infrastructure stages. Three affiliated bodies work to achieve this goal: CalCEF Ventures, an investment fund; CalCEF Innovations, leading analysis and product development; and CalCEF Catalyst, an industry acceleration platform. See calcef.org.

The **CalCEF Clean Energy Angel Fund** is an early-stage venture fund devoted to investments that transform the clean energy economy. See Crunchbase, "CalCEF Clean Energy Angel Fund," https://www.crunchbase.com/organization/calcef-clean-energy-angel-fund.

CalPERS, the California Public Employees' Retirement System, manages health benefits and pensions for public employees, retirees, and their families in California. It is the largest public pension fund in the US, with assets of over $354 billion at the start of 2018. See calpers.ca.gov.

Calysta converts methane into protein for seafood and livestock feed through the use of microorganisms. See calysta.com.

In a **cap and trade system** (also known as an **emissions trading system**), total emissions for a country or region are capped at some level; rights to emit within that cap are allocated to market participants, and then individual participants are allowed to sell unused rights to other parties that emit more than their permitted amounts. Typically, the total number of emission rights are reduced over time. The system encourages participants to cut their emissions to avoid having to purchase additional rights from other companies. The buying and selling of rights has the effect of making higher-emitting companies' output (e.g., electricity) more expensive, while making output from low-polluters cheaper. Cap and trade is an example of a market-based strategy (as opposed to a top-down, command-and-control system) that harnesses market forces to reduce emissions cost-effectively. See Union of Concerned Scientists, "Carbon Pricing 101," n.d., https://www.ucsusa.org/global-warming/reduce-emissions/cap-trade-carbon-tax. For details on the European Union's bloc-wide Emissions Trading System, see European Commission Climate Action, "The EU Emissions Trading System," 9 January 2018. https://ec.europa.eu/clima/policies/ets_en.

Carbon capture and sequestration (**CCS**), **carbon capture and storage**, and **carbon capture and use technology** (**CCUT**) refer to a set of technologies that prevents carbon dioxide (CO_2)—resulting from coal, oil, or natural gas combustion—from entering the atmosphere at its source. **Sequestration** involves transporting the gas via pipeline to depleted oil and gas wells or deep saline aquifers, and injecting it into porous rock formations, often a mile or more underground. Non-porous, impenetrable rock above the injection site prevents the gas from migrating upward (see U.S. Environmental Protection Agency, "Carbon Dioxide Capture and Sequestration: Overview," n.d., https://archive.epa.gov/epa/climatechange/carbon-dioxide-capture-and-sequestration-overview.html). **Use** involves making use of that CO_2 in industrial processes or converting it into biomass (see World Coal Association, "Carbon Capture, Use & Storage," 2015, http://www.worldcoal.org/reducing-co2-emissions/carbon-capture-use-storage).

Cement bond logging is the process of lining an oil or gas well with cement as it's being drilled to seal it off from the surrounding rock.

Chemical and Materials Enterprise Associates Capital. See **CMEA** below.

The U.S. **Clean Air Act** of 1963 was the first federal legislation requiring limits on air pollution to protect the health and property of citizens from the negative impacts of harmful air pollutants. Major amendments to this Act were passed in 1970 requiring that all dangerous stationary (power plants and large industrial facilities) and mobile (motor vehicles) sources be monitored and regulated; 1977 amendments strengthened emissions regulations on power plants; 1990 amendments added acid rain, ozone depletion, and air toxics (e.g., arsenic, mercury, lead, dioxin, benzene, etc.) to the scope. For more information on the Clean Air Act see Julie R. Domike and Alec C. Zacaroli (Editors), *The Clean Air Act Handbook* (Washington, D.C.: American Bar Association, 2013) and U.S. Environmental Protection Agency, "Carbon Pollution Standards: What EPA Is Doing," archived from original on 28 May 2015, https://web.archive.org/web/20150528004750/http://www2.epa.gov/carbon-pollution-standards/what-epa-doing.

Clean Energy Ministerial is a global forum that promotes policies and programs advancing clean energy technology. Its goal is to share best practices and insights and to encourage the transition to a global clean energy economy. The **Clean Energy Ministerial Dialogues** is an annual conference of energy ministers from the world's 24 major greenhouse gas-emitting nations, starting in 2010. The 2017 event was in Beijing, China, and the 2018 event will be jointly sponsored by the European Commission, Denmark, Finland, Norway, and Sweden. See www.cleanenergyministerial.org.

The **Climate Technology Centre and Network** (**CTCN**) promotes the accelerated transfer of environmentally sound technologies for low-carbon and climate-resilient development. The CTCN is run by the United Nations (specifically, the United Nations Framework Convention on Climate Change [UNFCCC] Technology Mechanism, the United Nations Environment Programme [UNEP], and the United Nations Industrial Development Organization [UNIDO]), and also draws on the climate technology experience of 11 other regional organizations. See www.ctc-n.org.

CMEA (**Chemical and Materials Enterprise Associates Capital**; now **Presidio Partners**) was an affiliate fund of New Enterprise Associates (NEA). CMEA was founded as a $40M fund with corporations and financial institutions as limited partners in 1989; it became independent in 1997. The firm's goal was to invest in startups focused on specialty chemicals and materials technologies; as Presidio Partners it has expanded its investment focus to include other sectors. See presidiopartners.com.

Concentrating solar power. See **solar thermal technology** below.

Copper indium gallium (di)selenide (**CIGS**) is a semiconductor material used to make thin film solar cells.

Corporate Average Fuel Economy standards. See **CAFE (Corporate Average Fuel Economy) standards** above.

D

The **Defense Advanced Research Projects Agency** (**DARPA**) fostered the research and development of computer networking and global positioning technologies that have since become commonplace features in consumer goods. DARPA is credited with enabling the development of innovations like the Global Positioning System (GPS), stealth fighter technology, and several generations of computer networking, among many others. Its mission is "to make pivotal investments in breakthrough technologies for national security." See www.darpa.mil.

Demand response programs offer financial incentives to electricity end-users to either reduce or shift their usage during times of peak demand. The programs allow utilities to forego building new generation facilities that are only needed for peak usage, and benefit the consumer by allowing utilities to avoid purchasing wholesale electricity at high, peak-demand prices, and avoiding rolling blackouts (see California Public Utilities Commission, "Demand Response," 2018, http://www.cpuc.ca.gov/General.aspx?id=5924). At the extreme, the Texas utility TXU Energy offers free overnight electricity to its customers, coupled with slightly higher daytime rates, making use of abundant nightly wind generation (see Clifford Krauss, and Diane Cardwell, "A Texas Utility Offers a Nighttime Special: Free Electricity," *The New York Times*, 8 November 2015, http://www.nytimes.com/2015/11/09/business/energy-environment/a-texas-utility-offers-a-nighttime-special-free-electricity.html).

John Doerr is a partner and chair of the venture capital firm, Kleiner Perkins Caufield & Byers and has funded tech companies such as Google, Amazon, and Intuit. He pushed KPCB into cleantech by creating a $500 million Green Growth Fund to support new energy technologies, and called clean energy the "largest economic opportunity of the 21st century" (Kent Garber, "John Doerr: Venture

Capitalist Pushes Green Technology," *U.S. News & World Report*, 22 October 2009, para. 3, https://www.usnews.com/news/best-leaders/articles/2009/10/22/john-doerr-venture-capitalist-pushes-green-technology).

Draper Fisher Jurvetson (**DFJ**). Tim Draper and John Fisher founded the venture capital firm in 1985, and Jurvetson's name was added to the door after a successful investment in Hotmail (acquired by Microsoft) in 1995. DFJ currently has nearly 800 investments, focused on three categories: consumer applications and services, enterprise infrastructure and applications, and disruptive technologies. See dfj.com.

E

ECOtality manufactured electric car chargers and operated Blink EV, a network of charging stations. Founded in 1999, the company filed for bankruptcy in 2013; its Blink network was sold to the Car Charging Group. See Jim Motavalli, "Electric Car Charger Company, ECOtality, Goes Bankrupt, Stranding 13,000 Docking Stations," *Huffington Post*, 13 October 2013, http://www.huffingtonpost.com/2013/10/13/electric-car-charger_n_4086326.html.

EERE (**Office of Energy Efficiency and Renewable Energy**) is responsible for developing and promoting energy efficiency and renewables technologies, as well as policies designed to stimulate their further development and use by the private sector. See http://energy.gov/eere/about-us.

Electromagnetic Systems Laboratory (**ESL**) was founded in Palo Alto, California in 1964 to provide reconnaissance and location-finding capabilities to the U.S. military in a way that enabled real-time integration of intelligence information. The company had its initial public offering on the Nasdaq exchange in the mid 1970s, and merged with TRW in 1978. In 2002, TRW—including ESL—was acquired by Northrup Grumman Corporation, a major defense intelligence contractor.

Emissions trading system. See **cap and trade system** above.

The **Energy Frontier Research Center** (**EFRC**) program was founded by the U.S. Department of Energy's Office of Science to accelerate the discovery of new breakthrough energy technologies. The program's 32 centers across the country consist of partnerships between universities, national labs, nonprofits, and for-profit companies conducting fundamental research focusing on "grand challenges" in the energy industry. See https://science.energy.gov/bes/efrc.

The **Energy Independence and Security Act of 2007** (**EISA**), mandated increases in vehicle fuel economy, set minimum production levels for domestic biofuels, and included significant energy efficiency provisions. Congress of the United States of America, *Energy Independence and Security Act of 2007*, H.R. 6, http://www.gpo.gov/fdsys/pkg/BILLS-110hr6enr/pdf/BILLS-110hr6enr.pdf.

Energy Innovation Hubs, under the Department of Energy, "bring together top researchers from academia, industry and the government laboratories with expertise that spans multiple scientific and engineering disciplines.... These teams orchestrate an integrated, multidisciplinary systems approach to overcoming critical technological barriers to transformative advances in energy technology." An example often used to describe the goal of these hubs is to create an environment for innovation like the one that existed at Bell Labs in 1947 when the transistor was invented. See https://science.energy.gov/bes/research/doe-energy-innovation-hubs/.

The **Energy Policy Act of 2005** was a multi-faceted law that covers energy efficiency, renewable energy, tax incentives, vehicle fuel economy, and the wholesale electric utility marketplace. Congress of the United States of America, *Energy Policy Act of 2005*, Public Law 109–58, http://www.gpo.gov/fdsys/pkg/PLAW-109publ58/pdf/PLAW-109publ58.pdf.

Energy Star was the first example of what has since become known as an appliance labeling program wherein appliances like refrigerators, ovens, and clothes dryers are evaluated, with the most efficient models receiving an "Energy Star" label. The voluntary program was established by the U.S. Environmental Protection Agency in 1992 under the Clean Air Act. See Energy Star, "Overview," n.d., https://www.energystar.gov/about. Research has shown that consumers are much more likely to buy appliances with this type of label than without one, and are willing to pay significantly more for them. See David O. Ward et al., "Factors Influencing Willingness-to-Pay for the ENERGY STAR® Label," *Energy Policy* 39, no. 3 (2011): 1450–1458, doi:10.1016/j.enpol.2010.12.017.

An **EPC contractor** (engineering, procurement, and construction) is an outside firm with the expertise and resources to build an industrial facility like a power station. The contractor becomes responsible for building the facility by a certain date and for a certain price, and guarantees that it will perform as designed (or else pay damages).

Exxon Chemical, short for ExxonMobil Chemical, manufactures and markets petrochemical products worldwide. See exxonmobilchemical.com.

F

Feed-in tariff is an energy supply policy that promotes the adoption of renewable energy. Feed-in tariffs are generally long-term contracts (5-20 years) offering a guarantee of payment to renewable producers; for rooftop solar, this guarantee would be a fixed price for all electricity produced by the installation and fed into the grid. These policies are widely used in Europe as well as several states in the US. The fixed price is generally above retail electricity prices. Electric utilities pay for renewable energy supplies at cost and roll the additional expense into the average rate charged to all their electricity consumers. This type of policy does encourage expanded use of all types of renewable energy, but can result in unequal costs among renewable energy sources and higher electricity rates if the cost of renewables is higher than that of conventional energy sources. See National Renewable Energy Laboratory, "Feed-In Tariffs," n.d., https://www.nrel.gov/technical-assistance/basics-tariffs.html; and U.S. Energy Information Agency, "Feed-In Tariff: A Policy Tool Encouraging Deployment of Renewable Electricity Technologies," 30 May 2013, https://www.eia.gov/todayinenergy/detail.cfm?id=11471.

The **financial crisis** broadly refers to the events that began in late 2007 by the collapse of the U.S. housing bubble. It turned out that a number of banks were (over)exposed to the housing market collapse through a variety of investments. Lehman Brothers—a storied investment bank—filed for bankruptcy, J.P. Morgan bought Bear Stearns, and Bank of America bought Merrill Lynch. During this time, as the various banks struggled to survive, they effectively stopped lending to or investing in renewable projects (or, really, anything else).

Fisker Automotive is a hybrid-electric sportscar company that was founded in 2007 and filed for bankruptcy in November 2013 due to operational difficulties; it was an early rival to Tesla Motors. It received $1.2 billion in venture capital investments prior to its bankruptcy and its assets were later bought by Chinese Wanxiang Group, which has started production of cars again. See Angela

Grelling Keane, "Fisker to Sell Assets in Bankruptcy," *Bloomberg*, 22 November 2013, http://www.bloomberg.com/news/articles/2013-11-22/fisker-to-sell-assets-in-bankruptcy-at-139-million-loss; and Paul Lienert, "Exclusive: China's Wanxiang to Rebrand Fisker as Elux – Sources," *Reuters*, 22 February 2015, https://www.reuters.com/article/us-autos-fisker/exclusive-chinas-wanxiang-to-rebrand-fisker-as-elux-sources-idUSKBN0LR00720150223.

Formation 8 is a San Francisco-headquartered venture capital firm. It was formed in 2012 to support startups looking to expand into the Asian market. The firm raised nearly $1 billion in two funds and announced in 2015 that it would not raise third fund (see Dan Primack, "Venture Firm Formation 8 Calls It Quits," *Fortune*, 5 November 2015, http://fortune.com/2015/11/05/venture-firm-formation-8-calls-it-quits/; and Dan Primack, "What Caused Formation 8 To Split, and What Comes Next?" *Fortune*, 10 November 2015, http://fortune.com/2015/11/10/what-caused-formation-8-to-split-and-what-comes-next/). See formation8.com and 8vc.com.

The **Fuels from Sunlight Hub** was one of the U.S. Department of Energy's Energy Innovation Hubs established to harness solar energy for the production of liquid transportation fuel. It's official name is now the **Joint Center for Artificial Photosynthesis** (**JCAP**). See U.S. Department of Energy, "Energy Department to Provide $75 Million for 'Fuels from Sunlight' Hub" (news release), 28 April 2015, https://energy.gov/articles/energy-department-provide-75-million-fuels-sunlight-hub.

G

GHG, short for "greenhouse gas."

The **Global CCS Institute** is dedicated to accelerating the adoption of carbon capture and storage (CCS) by gathering and sharing information with its diverse audience. Members include governments (UK, US, Canada, Australia, China, Japan), corporations (e.g., ExxonMobil, BHP, Kawasaki, Toshiba), and financial organizations (e.g., Asian Development Bank). The Global CSS Institute is financially independent and is based in Australia with offices in Washington DC, Brussels, Beijing, and Tokyo. See globalccsinstitute.com.

Al Gore was the 45th Vice President of the United States, from January 2003 until 2011. He prioritized the climate crisis during his time as Vice President and earlier as a U.S. Senator and Representative. He won the Nobel Peace Prize in 2007 with the Intergovernmental Panel on Climate Change for his work on the environment, and remains one of the leading figures in the global movement to solve the climate crisis (see Al Gore, "Climate Leadership," n.d., https://www.algore.com/about/the-climate-crisis). In 2006, he released the documentary, *An Inconvenient Truth* (see A. O. Scott, "Warning of Calamities and Hoping for a Change in 'An Inconvenient Truth,'" *The New York Times*, 24 May 2006, http://www.nytimes.com/movies/movie/342290/An-Inconvenient-Truth/awards). Gore became a partner with Kleiner Perkins Caufield & Byers in 2007 and also sits on Apple's board of directors; he is the founder of the Alliance for Climate Protection (also known as the Climate Reality Project).

H

Haloid Corporation see **Xerox Corporation** below.

Hambrecht & Quist (**H&Q**, now **WR Hambrecht**) is a Silicon Valley-based investment banking firm specializing in high-tech investing. It was purchased by Chase in 1999. See wrhambrecht.com.

A **heat pump** heats a building by taking advantage of the differential between inside and outside temperatures. It uses a working fluid that is cooler than the outside temperature, which absorbs heat from the outside air. Once the fluid is inside, the pump compresses it so that it becomes warmer than the inside temperature, thereby transferring heat to the inside of the building. A refrigerator operates using the opposite sequence of transformations and, indeed, the heat pump's operation can be reversed to air condition the building in the summer.

Dr. Stefan Heck is an energy business strategist, author, and entrepreneur and former energy program director at McKinsey. He is a Co-Founder and CEO of NAUTO, a company focused on building an information network to improve urban transportation mobility sustainably. Heck is a Consulting Professor at the Precourt Institute for Energy at Stanford University and a research fellow at the Steyer-Taylor Center for Energy Policy and Finance at Stanford.

Heliotrope Technologies, borne out of Lawrence Berkeley National Labs, has developed a coating for window glass that can modulate both the amount of light and the amount of heat that comes into a room. The inventor is Jill Fuss, a professor in the Lab's nanotechnology group. See heliotropetech.com.

HelioVolt was Texas-based U.S. solar energy and manufacturing company founded in 2001; the firm went out of business in 2014. See Eric Wesoff, "Solar Grim Reaper Alert: CIGS Aspirant HelioVolt Gives Up the Ghost," *Greentech Media*, 25 February 2014, https://www.greentechmedia.com/articles/read/Solar-Grim-Reaper-Alert-CIGS-Aspirant-HelioVolt-Gives-Up-the-Ghost.

I

In-Q-Tel is the nonprofit venture capital arm of the U.S. Central Intelligence Agency, investing in cutting-edge technology of interest to the intelligence community. See Rick E. Yannuzzi, "In-Q-Tel: A New Partnership Between the CIA and the Private Sector," Central Intelligence Agency, 2000, https://www.cia.gov/library/publications/intelligence-history/in-q-tel; and www.iqt.org.

The **Intergovernmental Panel on Climate Change** (**IPCC**) was established by the United Nations Environment Programme (UNEP) and the World Meteorological Organization in 1988. It reviews and assesses the most recent scientific, technical, and socioeconomic research related to the understanding of climate change. See ipcc.ch.

The **International Energy Agency** (**IEA**) is an autonomous organization that provides energy data and analysis for its 29 member countries, with a focus on ensuring reliable, affordable, clean energy. Its main areas of focus are energy security, economic development, environmental awareness, and engagement. See www.iea.org.

The **International Renewable Energy Agency** (**IRENA**) is an intergovernmental agency established in January 2009 that promotes the spread of renewable energy around the world, with a focus on enabling sustainable development. See irena.org.

The **International Technology Roadmap for Semiconductors** (**ITRS**) is the 15-year assessment of the semiconductor industry's future technology requirements; these future needs drive present-day strategies for worldwide R&D among manufacturers, universities, and national labs. The objective of the ITRS is to ensure cost-effective advancements in the performance of the integrated circuit, which in turn continue the health and success of the industry. See www.itrs2.net.

The **Investment Tax Credit** (**ITC**) is a 30% federal tax credit on the cost of installing a new solar system. A hypothetical solar system that costs $10,000 to install would result in a $3,000 tax credit for the system's owner (see Solar Energy Industries Association, "The Solar Investment Tax Credit (ITC)," 27 January 2015, http://www.seia.org/sites/default/files/ITC%20101%20Fact%20Sheet%20-%201-27-15.pdf). The ITC, as well as the Production Tax Credit (PTC) for wind energy, were extended for 5 years by U.S. Congress in December 2015; solar projects started after 2018 will have incrementally smaller credits, leveling off at 10% in 2022 (see Daniel Cusick, "Renewables Boom Expected Thanks to Tax Credit," *Scientific American*, 21 December 2015, http://www.scientificamerican.com/article/renewables-boom-expected-thanks-to-tax-credit/).

An **investment thesis** guides a venture capital firm's investment decisions. The thesis can focus on a particular industry, geography, or funding stage (e.g., seed, Series A-C, growth).

An **isothermal** process is one in which the temperature remains constant.

Itron is a global technology and services company focused on energy and water. The firm manufactures a variety of meters for use by electric, gas, and water utilities to help them manage energy resources efficiently and responsibly. See itron.com.

Ivanpah is a 392 MW solar power tower project located in the Mojave Desert of California. Ivanpah was BrightSource Energy's first completed project, and the first time a solar power tower was built in the U.S. since the 1980s. See ivanpahsolar.com.

J

K

Jay D. Keasling is a well-known and highly regarded chemical- and bio-engineering professor at the University of California, Berkeley, specializing in metabolic engineering for advanced biofuel production and drug delivery. He also serves as Associate Laboratory Director (Biosciences) at the Lawrence Berkeley

National Laboratory, is chief executive officer of the Joint BioEnergy Institute, and was a founder of biofuel cleantech startup Amyris Biotechnologies.

Vinod Khosla founded Khosla Ventures in 2004, which has invested in cleantech technologies such as wood-based biofuel, batteries, and water purification. He previously spent 20 years investing at KPCB. Before his investing career, he co-founded Sun Microsystems in 1982. See "The World's Billionaires: #1198 Vinod Khosla," *Forbes*, 14 April 2016, http://www.forbes.com/profile/vinod-khosla/.

Kleiner Perkins Caufield & Byers (**KPCB**) was founded in 1972 and has since invested in hundreds of companies, including Google, Amazon, Sun Microsystems, Netscape, AOL, and Electronic Arts. KPCB began to focus on cleantech, or "greentech" as the firm termed it, shortly after the dot-com bubble burst. They invested in renewable fuels, energy generation, energy storage, and sustainable agriculture. See kpcb.com.

L

K. V. Lakshmi is an Associate Professor at Rensselaer Polytechnic Institute and director of the Baruch '60 Center for Biochemical Solar Energy Research. See https://science.rpi.edu/chemistry/faculty/k-v-lakshmi.

Lawrence Berkeley National Laboratory (**LBNL**) is a member of the national laboratory system supported by the U.S. Department of Energy through its Office of Science. It is managed by the University of California and conducts unclassified research across a range of scientific disciplines. The lab was founded in 1931 by Ernest Orlando Lawrence, a UC Berkeley physicist who won the 1939 Nobel Prize in physics for his invention of the cyclotron. See lbl.gov.

Lehman Brothers. See **financial crisis** above.

LUZ II was an Israeli company that built nine solar thermal projects totaling 354 MW in the United States between 1984 and 1990. It changed its name to BrightSource Energy in 2006. LUZ's projects were named SEGS 1 through 9 and are collectively known in the industry as "the SEGS projects." See California Energy Commission, "Solar Energy Projects in California," 2018, http://www.energy.ca.gov/sitingcases/solar/.

M

Materials science is an interdisciplinary subject, spanning the physics and chemistry of matter, engineering applications, and industrial manufacturing processes. Materials science, investigating the structures and properties materials at nano-, micro-, and macro-scales, includes developments that span nanotechnology, quantum computing, and nuclear fusion, as well as advanced medical technologies.

Merchant is an industry term referring to a power plant that is not selling output under a fixed-price, long-term contract (a PPA), but rather is selling at whatever price prevails in the market at the time of sale.

MIT's "Rad Lab" refers to the MIT Radiation Laboratory, which operated from 1940 to 1945 during World War II. Rad Lab made contributions to the development of microwave radar technology in support of the war effort. See https://www.ll.mit.edu//about/History/RadLab.html.

Moore's Law is an observation and prediction that the number of transistors in an integrated circuit doubles every two years. The theory has been reinforced over the years via the numerous constant innovations by scientists and engineers.

Elon Musk is a serial entrepreneur known for founding or playing a significant role in the creation of PayPal, SpaceX, Tesla Motors, and SolarCity. As of 2018, he serves as CEO and and lead designer of SpaceX; CEO and product architect at Tesla Motors; CEO of Neuralink, which is developing brain–machine interfaces; co-founder and Chairman of the nonprofit OpenAI; and founder and CEO of The Boring Company, developing advanced technologies for building tunnels.

N

National Renewable Energy Laboratory (**NREL**) is the United States' primary laboratory for renewable energy and energy efficiency research. The Colorado-based organization is government-owned, contractor-operated, and funded by the Department of Energy. See nrel.gov.

Nest Labs designs Wi-Fi enabled and sensor-driven programmable thermostats and smoke detectors; it was acquired by Google in 2014 for $3.2 billion. See nest.com.

New Enterprise Associates (NEA) is a global venture capital firm that invests in technology and healthcare at all stages of company growth. See nea.com.

Non-recourse debt is collateralized by specific assets but not the underlying creditworthiness of the borrower. If the cash flow from the assets used as collateral is insufficient to repay the loan, the borrower is not obligated to pay the debt and the lender has no recourse to recover their investment.

NREL. See **National Renewable Energy Laboratory** above.

Nth Power is a San Francisco-based venture capital firm founded by Nancy Floyd that invests in energy technology companies. See nthpower.com.

O

Office of Energy Efficiency & Renewable Energy. See **EERE** above.

The **Offset Strategy** was a Cold War strategy to counter the tremendous advantage the Soviet Union had gained in military hardware (missiles, boats, and airplanes) by using smarter systems incorporating advanced digital sensor and control technologies like GPS, the Internet, laser guidance, and so on. See Defense Science Board Task Force, *The Roles and Authorities of the Director of Defense Research and Engineering* (Office of the Secretary of Defense for Acquisition, Technology, and Logistics, 2005), iv, https://www.acq.osd.mil/dsb/reports/2000s/ADA440086.pdf.

Opower is a SaaS company focused on customer engagement and energy efficiency, providing tools for consumers and utilities to make more informed energy decisions; it was acquired by Oracle in 2016. See opower.com.

P

PCAST. See **President's Council of Advisors on Science and Technology** below.

A **phasor measurement unit** (PMU) is a device that can measure instantaneous voltage, current, and frequency at a specific location in the power grid, sampling data over 1,200 times per second. Multiple units spread over the grid can be synchronized to create a high-resolution "picture" of conditions throughout the grid. See U.S. Department of Energy, *Synchrophasor Technologies and Their Deployment in the Recovery Act Smart Grid Programs*, August 2013, 2–3, https://www.smartgrid.gov/files/Synchrophasor_Report_08_09_2013_DOE_2_version_0.pdf.

Polysilicon (**polycrystalline silicon**) is a highly purified, crystalline form of silicon, used an element that has semiconductor-like material properties. It is used as raw material in the manufacture of solar photovoltaic cells. See AE Polysilicon, "What is Polysilicon?" 2016, http://www.aepolysilicon.com/Solar-Education/Polysilicon/.

Power electronics control how electrical power is controlled and converted—between different currents, voltage levels, and frequencies—throughout the grid. About 30% of electricity in the United States flows through various types of power converters, projected to rise to 80% in the coming decades. Making power conversion more efficient, particularly with distributed solar energy, is vital. See U.S. Department of Energy, ARPA-E, "ADEPT: Agile Delivery of Electrical Power Technology," 12 July 2010, http://arpa-e.energy.gov/?q=arpa-e-programs/adept.

A **power purchase agreement** (**PPA**) is a long-term contract (20-25 years) for the purchase of electricity from a renewable source (like solar), typically bundled together with renewable energy tax credits, at a fixed price. In solar power PPAs—called the "solar services" model—a third-party developer owns, operates, and maintains the photovoltaic system, and a host customer agrees to site the system on its property and purchases the system's electric output from the services provider for a predetermined period. The host customer receives stable, and sometimes lower-cost electricity, and the solar services provider receives tax credits as well as income generated from the sale of electricity to the host customer. See U.S. Environmental Protection Agency, "Solar Power Purchase Agreements," n.d., https://www.epa.gov/greenpower/solar-power-purchase-agreements.

The **President's Council of Advisors on Science and Technology** (**PCAST**) was created by President Obama in 2009. It is an advisory group of leading scientists and engineers who directly advise the president and make policy recommendations. See https://obamawhitehouse.archives.gov/administration/eop/ostp/pcast.

Presidio Partners. See **CMEA** above.

The **Production Tax Credit** (**PTC**) is the U.S. wind industry's most important tax initiative, providing a 2.3-cent per kilowatt hour (kWh) credit for the first 10 years of a facility's operation. The PTC, as well as the solar energy Investment Tax Credit (ITC), were extended for 5 years by U.S. Congress in December 2015. The PTC will step down over the 5 years of the legislation to 40% of its 2016 level by 2020. See Daniel Cusick, "Renewables Boom Expected Thanks to Tax Credit," *Scientific American*, 21 December 2015, http://www.scientificamerican.com/article/renewables-boom-expected-thanks-to-tax-credit/; and Union of Concerned Scientists, "Production Tax Credit for Renewable Energy," n.d., https://www.ucsusa.org/clean-energy/increase-renewable-energy/production-tax-credit.

Project financing is a method of funding a large, long-term infrastructure project in which the debt raised is repaid from the cash flow generated by the project after it is completed. Project-financed debt is non-recourse and it remains off the company's balance sheet, and thus does not affect its credit rating or its ability to borrow for other business purposes.

Q

R

R&D, **RD&D**, and **RDD&D** refer to the following, in this order: research, development, demonstration, and deployment.

A **renewable portfolio standard** (**RPS**), also known as a **renewable electricity standard**, is a mandate for utilities to have a minimum percentage of their electricity come from renewable sources, such as solar, wind, biomass, and hydroelectric. The goal is to increase the production of energy from renewable sources; such a program works most effectively in conjunction with tax credits such as the Investment Tax Credit and Production Tax Credit. See National Renewable Energy Laboratory (NREL), "Renewable Portfolio Standards," n.d., para.1, https://www.nrel.gov/technical-assistance/basics-portfolio-standards.html; Cheryl Cox and Amanda Singh (Editors), *Renewables Portfolio Standard Annual Report November 2017*, California Public Utilities Commission, http://www.cpuc.ca.gov/uploadedFiles/CPUC_Website/Content/Utilities_and_Industries/Energy/Reports_and_White_Papers/Nov%202017%20-%20RPS%20Annual%20Report.pdf.

A **revenue-neutral carbon tax** has been advocated by former Secretary of State George Shultz and Nobel Laureate Gary Becker. They proposed a tax on carbon, combined with the elimination of all forms of government subsidy throughout the energy economy in order to level the playing field between different energy sources, and in so doing reflect the environmental cost of each alternative. George P. Shultz and Gary S. Becker, "Why We Support a Revenue-Neutral Carbon Tax," *The Wall Street Journal*, 7 April 2013, http://www.wsj.com/articles/SB10001424127887323611604578396401965799658.

S

SCADA systems. See **supervisory control and data acquisition systems** below.

Lee Schipper, an energy scholar and iconoclast, had tenures at Stanford University, the University of California, Berkeley, Lawrence Berkeley National Labs, and the International Energy Agency in Paris. His 100+ publications during his career focused primarily on energy economics and transportation. See Stanford University, "In Memoriam, Lee Schipper," http://peec.stanford.edu/people/profiles/Lee_Schipper.php.

SEGS. See **solar energy generating systems** below.

SEMATECH was founded in 1987 by a consortium of 14 large U.S.-based semiconductor manufacturers in collaboration with the U.S. government to help them compete with the Japanese semiconductor industry. The U.S. Department of Defense provided $500 million of cooperative funding to SEMATECH via the Defense Advanced Research Projects Agency (DARPA), spread over nearly a decade. In 1996, the organization decided to eliminate U.S. government funding and to expand its membership to include key major non-U.S. semiconductor manufacturers. SEMATECH members currently represent about half of the worldwide chip production and continue to collaborate on pre-commercial semiconductor research and technology development projects. See Robert D. Hof, "Lessons from SEMATECH," *MIT Technology Review*, 25 July 2011, https://www.technologyreview.com/s/424786/lessons-from-sematech/; and Larry Browning, Janice M. Bayer, and Judy C. Shelter, "Building Cooperation in a Competitive Industry: SEMATECH and the Semiconductor Industry," *Academy of Management Journal*, 38 no. 11 (1995), 112–151.

Seraph Capital Forum is a network of women executives, typically with experience across a wide variety of industries, investing alongside each other. See Crunchbase, "Seraph Capital Forum," 2018, https://www.crunchbase.com/organization/seraph-capital-forum.

Silicon Energy MN was a designer and manufacturer of photovoltaic (PV) modules; it was bought by Itron in 2003. See Dan Gallagher, "Silicon Energy Acquired for $71.2 Million," *San Francisco Business Times*, 21 January 2003, http://www.bizjournals.com/eastbay/stories/2003/01/20/daily14.html?page=all.

Silicon Graphics is a manufacturer of computer software and hardware, including high-performance computing solutions, servers for datacenter deployment, and visualization products. Founded in the 1980s, it later went through bankruptcy and was acquired in 2009 by Rackable Systems and later purchased by Hewlett Packard Enterprise (HPE) in 2016. See JohathanVanian, "Hewlett Packard Enterprise Just Spent Millions on This Supercomputer Maker," *Fortune*, 11 August 2016, http://fortune.com/2016/08/11/hewlett-packard-enterprise-sgi-supercomputer/.

Silicon nanowires have a wide range of applications including photovoltaics, lithium batteries, and sensors. For technical detail on silicon nanowires, see Sigma Aldrich, "Silicon Nanowires," *Material Matters* 6 (2011): 1, http://www.sigmaaldrich.com/content/dam/sigma-aldrich/articles/material-matters/pdf/silicon-nanowires.pdf.

Smart grid technology refers to various electronic components built into a modern utility grid allowing two-way communication between customer and utility, as well as sensors throughout the grid providing data to the utility. These hardware components, along with sophisticated software, work together to respond quickly to changes in electricity demand. See U.S. Department of Energy Office of Electricity Delivery and Energy Reliability, "What Is the Smart Grid?" n.d., https://www.smartgrid.gov/the_smart_grid/smart_grid.html.

SolarCity is the largest solar power provider in the US, serving homeowners, government, nonprofits, and businesses. It was among the first rooftop solar companies to go public (in 2013). In 2016, Tesla Motors acquired SolarCity for $2.6 billion. Elon Musk is the company's chairman and largest shareholder. See solarcity.com.

Solar energy generating systems (**SEGS**) refers to a series of nine solar thermal projects by the Israeli company LUZ, collectively referred to as "the SEGS projects." See California Energy Commission, "Solar Energy Projects in California," 2018, http://www.energy.ca.gov/sitingcases/solar/.

Solar thermal technology (also referred to as **concentrating solar power**) collects sunlight to heat water or air. In the electric utility industry, there are two dominant technologies: trough and power tower. The trough design uses sunlight to heat fluid that is used to operate a steam turbine. The tower design uses a tall tower located in the center of rings of mirrors; the mirrors concentrate sunlight to heat fluid that is used to operate a steam turbine. This technology was originally commercialized by LUZ, an Israeli company that built a series of such plants in the United States in the 1980s (see SEGS above).

Solyndra was a photovoltaic solar company that was founded in 2005 and went bankrupt in October 2012 when the price of solar panel manufacturing fell worldwide. It received a $535 million loan guarantee from the Department of Energy that was highly publicized (see Washington Post, "A History of Solyndra," 13 September 2011, https://www.washingtonpost.com/politics/a-history-of-solyndra/2011/09/13/gIQA1r5qQK_story.html). Its bankruptcy and shutdown resulted in widespread criticism of the U.S. Department of Energy's loan guarantee program (see U.S. House of Representatives, Energy and Commerce Committee, "Committee Releases Extensive Report Detailing Findings of Solyndra Saga," 2 August 2012, https://energycommerce.house.gov/news/committee-releases-extensive-report-detailing-findings-solyndra-saga/).

SunPower, founded in 1985, is a leading designer and producer of high-efficiency photovoltaic cells, solar panels, and roof tiles. The company is listed on Nasdaq; French oil and gas energy giant Total S.A. owns almost 65% of the outstanding shares. See www.sunpower.com.

The **SunShot Initiative** was announced by the Solar Technologies Office at the U.S. Department of Energy in February 2011 with the goal of reducing the cost of solar energy to be cost-competitive with fossil fuel-generated electricity in most places in the United States by 2020—a 75% decrease from 2011 costs. See U.S. Department of Energy, Office of Energy Efficiency & Renewable Energy, "The Sunshot Initiative," n.d., https://energy.gov/eere/solar/sunshot-initiative.

Supervisory control and data acquisition (SCADA) systems use computers and communications networks to gather field data from numerous remote locations, perform numerical analysis, and generate trends and summary reports. In their most advanced form, these systems can be used to control equipment located in the field, such as over a smart grid. SCADA systems make it possible to control a process that is distributed over a large area, with just a small group of people located in a single room, and are key components of a reliable electricity supply. See Keith Stouffer, Joe Falco, and Karen Kent, *Guide to Supervisory Control and Data Acquisition (SCADA) and Industrial Control Systems Security*, National Institute of Standards and Technology, 2007, Section 2.1, https://www.dhs.gov/sites/default/files/publications/csd-nist-guidetosupervisoryanddataccquisition-scadaandindustrialcontrolsystemssecurity-2007.pdf.

Symyx Technologies was a company that specialized in informatics and automation products and whose software solutions were used in scientific research. Symyx merged with Accelrys in 2010, and the combined firm was purchased by Dassault Systems in 2014 and renamed BIOVIA. See www.accelrys.com.

T

Tax equity is a passive ownership interest in an asset, where the investor receives a return based on cash flow from the asset as well as income tax credits and benefits. Tax equity investors usually have a large tax obligation, and use, for example, a solar power installation to reduce future tax liabilities. See U.S. Partnership for Renewable Energy Finance, "U.S. Renewable Energy Tax Equity Investment and the Treasury Cash Grant Program," June 2011, http://www.uspref.org/white-papers/30-u-s-renewable-energy-tax-equity-investment-and-the-treasury-cash-grant-program.

Taxon Biosciences is a microbiome company focused on discovering and commercializing microbiomes for use in agriculture, life sciences, and oil and gas sectors. The company was acquired by DuPont in 2015. See taxon.com.

Technology Readiness Levels are metrics originally developed by NASA and the Department of Defense that are now used by other agencies, including the Department of Energy. There are nine levels of readiness, ranging from TRL 1 (Basic principles observed and reported) through TRL 9 (Actual system proven through successful mission operations). The use of TRLs allows for a consistent yardstick to measure a variety of technologies and projects. See U.S. Department of Defense, Technology Readiness Assessment (TRA) Guidance, 13 May 2011, 2-13 and 2-14, http://www.acq.osd.mil/chieftechnologist/publications/docs/TRA2011.pdf.

Tesla Motors designs, manufactures, and sells electric vehicles and electric vehicle powertrain components. It maintains a network of supercharging stations across the US and in Europe and Asia. As of late 2017, the company had sold over 250,000 cars worldwide; 2016 saw the introduction of its Model X crossover vehicle and 2017 saw the initial production of the mass-market, affordable Model 3. Tesla was founded in 2003 and is led by CEO Elon Musk; it went public in 2010. See teslamotors.com/about.

U

United Nations Framework Convention on Climate Change (UNFCCC) is the framework established for international climate policy negotiations at the 1992 Rio Earth Summit. One requirement of the framework is for all participating nations to meet at least once a year at a Conference of Parties (COP) to the convention to discuss progress and plans. COP-23 took place in Bonn, Germany in late November 2017. See unfccc.int/2860.php.

U.S.–China Clean Energy Research Centers (CERC) facilitates collaboration between researchers in the United States and China to accelerate research, development, and deployment of clean energy technologies. See www.us-china-cerc.org.

V

The **Valley of Death** is a common term in the startup world, referring to the commercialization funding gap. Companies typically have a negative cash-flow period between its receipt of investor capital—when its technology is still in development—and when it begins generating customer revenue. See Martin Zwilling, "10 Ways For Startups To Survive The Valley Of Death," *Forbes*, 18 February 2013, http://www.forbes.com/sites/martinzwilling/2013/02/18/10-ways-for-startups-to-survive-the-valley-of-death/.

VantagePoint Capital Partners is a venture fund specializing in energy innovation and efficiency investments. See vpcp.com.

W

X

Xerox Corporation, formerly known as the **Haloid Corporation**, produces and sells printers, photocopiers, printing presses, and related supplies and services. See xerox.com.

Y

Z

Zilog supplies application-specific, embedded system-on-chip (SoC) solutions for the industrial and consumer markets. Its expertise includes SoCs, single board computers, application-specific software stacks, and development tools. The company's leading product for many years was the Z8 microprocessor, a novel and sophisticated 8-bit single-chip device. See zilog.com.

www.ingramcontent.com/pod-product-compliance
Lightning Source LLC
Chambersburg PA
CBHW060556200326
41521CB00007B/589